NUCLEAR STRUCTURE

INVESTIGATIONS IN PHYSICS
Edited by EUGENE WIGNER and ROBERT HOFSTADTER

1. Ferroelectricity by E. T. JAYNES
2. Mathematical Foundations of Quantum Mechanics by JOHN VON NEUMANN
3. Shell Theory of the Nucleus by EUGENE FEENBERG
4. Angular Momentum in Quantum Mechanics by A. R. EDMONDS
5. The Neutrino by JAMES S. ALLEN (Spring 1958)
6. Photonuclear Reactions by L. KATZ, R. SKINNER, and A. S. PENFOLD (to be published)
7. Radiation Damage by D. BILLINGTON and J. CRAWFORD (to be published)
8. Nuclear Structure by LEONARD EISENBUD and EUGENE P. WIGNER

NUCLEAR STRUCTURE

BY

Leonard Eisenbud and Eugene P. Wigner

PRINCETON, NEW JERSEY
PRINCETON UNIVERSITY PRESS
1958

Copyright © 1958, by Princeton University Press
London: Oxford University Press
L. C. Card 58-8562

Printed in the United States of America by
The William Byrd Press, Richmond, Virginia

PREFACE

Theoretical nuclear physics has at present a somewhat amorphous character. However, its state is far from stationary and one may reasonably hope for considerable crystallization in the near future.

It is easy to appreciate the reasons for the present formlessness of nuclear theory. Internucleon forces are not yet completely known and it is clear that they have a complex character. Even the consequences of a simple interaction are difficult to obtain for a system containing a large but finite number of particles. A good deal of effort has been expended, therefore, in the search for simple models in terms of which the broad regularities satisfied by nuclei could be understood. This search has led to a number of interesting but only partially successful models; these have proved very fruitful for the stimulation of experimental research, and for the development of further ideas on nuclear structure. One can hope that future investigations will clarify the limitations of these models and provide an understanding of the validity of different models for different groups of phenomena.

An exhaustive analysis of nuclear theory would require a very large volume. We have not attempted to equal the standards or the comprehensiveness of Blatt and Weisskopf's *Theoretical Nuclear Physics* or Sachs' *Nuclear Theory*. Instead, we have tried only to acquaint the reader with the points of view which we consider useful and interesting for the analysis of large groups of nuclear phenomena and among which he may wish to choose one or more for detailed study. However, neither the mathematical foundations of the points of view described, nor the empirical facts which support them, are developed in full detail. Even a thorough survey of the literature would have occupied more space than the whole of the present treatise, and we must ask the indulgence of all those whose work we failed to appreciate or understand, or whose work was left unmentioned because of the limitations in space. The emphasis is on the description and comparison of the assumptions of various theories, together with a discussion of only some of the consequences of these assumptions. For the most part the treatment is non-mathematical.

Since nuclear physics is developing so rapidly, a description of the subject must appear quickly if it is to be useful to its readers. We are indebted to the Princeton University Press for its efforts to give this book rapid publication. The material was originally prepared as a section for the large *Handbook of Physics* (McGraw-Hill) which is being edited by E. U. Condon. It appeared that the Handbook would require a considerable publication time so that a separate treatment of our work seemed advisable. Through an arrangement between the McGraw-

Hill Book Company and the Princeton University Press this book will also appear in the *Handbook of Physics*.

One of us (L.E.) wishes to take this opportunity to express his appreciation for the support of the joint Program of the Office of Naval Research and the U.S. Atomic Energy Commission while working on this book.

Leonard Eisenbud and Eugene P. Wigner

TABLE OF CONTENTS

Chapter 1.	General Features of Nuclei	3
1.1.	Nuclear Composition.	3
1.2.	Nuclear Masses: Binding Energies	4
1.3.	Types of Nuclear Instability: Spontaneous and Induced Transformations	6
	a. Natural α-radioactivity; Fission	7
	b. γ-radiation; Particle Emission	8
	c. β-decay	9
Chapter 2.	Systematics of Stable Nuclei; Details of Binding Energy Surfaces	12
Chapter 3.	Properties of Nuclear States; Ground States	17
3.1.	Spins and Moments	17
3.2.	The Size of Nuclei	20
Chapter 4.	Survey of Nuclear Reactions	24
4.1.	Types of Reaction, Cross-Sections, Excitation Functions	24
4.2.	Resonance Processes	26
4.3.	Direct Processes	28
4.4.	Table of the Most Important Reactions	29
Chapter 5.	Two-body Systems: Interactions between Nucleons	32
5.1.	Internucleon Forces	32
5.2.	Saturation Properties and Internucleon Forces	35
5.3.	Charge Independence of Nuclear Forces; The Isotopic or Isobaric Spin Quantum Number	36
Chapter 6.	Nuclear Models A: The Uniform Model	40
6.1.	General Remarks	40
6.2.	Powder and Shell Models	41
6.3.	Supermultiplet Theory	42
Chapter 7.	Nuclear Models B: Independent Particle Models	47
7.1.	General Features of the Independent Particle or Shell Models	47
7.2.	The L-S coupling Shell Model	49
7.3.	Comparison of the L-S and j-j Shell Models	50
7.4.	The j-j Coupling Shell Model	52
7.5.	Coupling Rules for the j-j Model	55
7.6.	Normal States and Low-Excited States	57
7.7.	Magnetic and Quadrupole Moments	59
7.8.	Problems of the j-j Model	60

CHAPTER 8. Nuclear Models, C: Many Particle Models   65
   8.1. The α-particle Model   65
   8.2. Collective Model   65
   8.3. Comparison of the j-j and the Collective Models. . . .   68

CHAPTER 9. Nuclear Reactions A: Close Collisions   71
   9.1. The Collision Matrix   71
   9.2. Qualitative Discussion of Resonance Phenomena . . .   75
   9.3. Derivation of the Resonance Formula   77
   9.4. Dependence of the Parameters on the Size of the Internal Region   80
   9.5. Radioactivity   81
   9.6. The Clouded Crystal Ball Model   82
   9.7. The Intermediate Coupling or Giant Resonance Model   84

CHAPTER 10. Nuclear Reactions B: Surface Reactions   89
   10.1. Angular Distribution in Stripping Reactions   90
   10.2. Electric Excitation.   94

CHAPTER 11. Interaction with Electron-Neutrino Fields   97
   11.1. Theory of β-decay   97
   11.2. Allowed and Forbidden Transitions   100
   11.3. Shape of the Spectrum   101
   11.4. Total Transition Probability   102

CHAPTER 12. Electromagnetic Transitions in Complex Nuclei .   105
   12.1. Introduction   105
   12.2. Radiative Transitions   106
   12.3. Single Particle Matrix Elements   109

References   112

Index   122

NUCLEAR STRUCTURE

CHAPTER 1
General Features of Nuclei

1.1. Nuclear Composition

The concept of the atomic nucleus was first advanced by Rutherford (1911) who showed that the positive charge and all but a small fraction of the total mass of an atom is concentrated in a central nuclear core with radius of the order of 10^{-5} that of the atomic radius. It was soon recognized that the charge on the nucleus of an atom, Ze, is an integral multiple of the electronic charge ($-e$ = electronic charge), and that the integer Z specifies the position of the atom in the periodic table. J. J. Thomson showed (1913) that the mass of a nucleus is not determined by its charge; nuclei of the same charge but different mass are called *isotopes*. On a mass scale defined by assigning the value 16 to the mass of the most abundant oxygen isotope the masses of all nuclei are close to integers. The integer closest to the mass of a nucleus on this scale is called the mass number, A, of the nucleus.

Modern theories of the nucleus stem from the discovery of the neutron —an uncharged particle of mass approximately that of the hydrogen atom (the mass number is unity)—by Chadwick (1932), and the suggestion by Heisenberg shortly thereafter that the elementary constituents of nuclei are neutrons and protons. A nucleus of charge number Z and mass number A is composed of Z protons and $N = A - Z$ neutrons. *Isotopes* are nuclei of equal Z but unequal N. The atoms formed on isotopic nuclei have practically identical chemical properties. Nuclei of equal N but unequal Z are called *isotones*. The elementary nuclear constituents are often referred to indiscriminately as *nucleons*. The number of nucleons in a nucleus is the mass number. Nuclei composed of the same numbers of nucleons (equal A) are referred to as *isobars*. There is considerable evidence for at least a rough equivalence of the properties of the neutrons and protons within nuclei. As a consequence the isobars play a role in the physics of nuclei comparable to the role of isotopes in atomic physics. For the designation of a particular nuclear species among isobars of given A the number $T_\zeta = \frac{1}{2}(N - Z)$, which measures the excess of the number of neutrons over the number of protons in the nucleus, has considerable theoretical usefulness.

Clearly any pair of the four numbers A, Z, N, T_ζ may be used to characterize the composition of a nucleus. The customary symbolic representation employs the numbers Z, A; the number Z is given by the chemical symbol of the element while A appears as a superscript.

Thus the nucleus with $Z = 79$, $A = 197$ is designated by Au^{197}. Sometimes the charge number is explicitly given as a subscript: Au^{197}_{79}.

The ranges of Z, A for known nuclei are $Z = 0$ (neutron) to $Z = 101$; $A = 1$ (proton, neutron) to $A = 256$.

The properties of the fundamental nuclear constituents are tabulated below:

	Mass	Charge	Spin	Magnetic Moment in Nuclear Magnetons	Statistics
neutron	1.00898	0	$\frac{1}{2}$	-1.9135	Fermi
proton	1.00759	e	$\frac{1}{2}$	2.7928	Fermi

Masses are given in the $O^{16} = 16$ scale. Magnetic moments are expressed in nuclear magnetons ($= e\hbar/2M_p c$, where M_p is the proton mass). The usual tables of atomic masses give the mass of the atom, i.e., the nuclear mass augmented by the mass of Z electrons or $548 \times 10^{-6} Z$.

The N, Z values for the known stable nuclei are presented in the "Segré Chart" (see Fig. 2.1). Isotopes are grouped along horizontal lines in the chart; isotones, along vertical lines. We shall return to a discussion of the significance of the structure of this chart in Chapter 2.

1.2. Nuclear Masses: Binding Energies

The energy required to decompose a nucleus into its constituent nucleons is called the *binding energy*, B, of the nucleus. The binding energy is related, through Einstein's mass-energy equation, to the difference between the mass of the nucleus and the sum of the masses of the constituent nucleons. If ΔM is this mass difference—the so-called mass defect of the nucleus—then $B = \Delta M c^2$ (where c is the velocity of light). Thus, if $M(Z, N)$ is the mass of a neutral atom with Z protons and N neutrons, and M_H and M_n are the masses of the neutral hydrogen atom and of the neutron, the binding energy is given by:

$$B(Z, N) = [M_H Z + M_n N - M(Z, N)]c^2.$$

(The mass of the electrons in the neutral atom is taken into account through the use of M_H, the mass of the hydrogen atom, rather than the proton mass). It is customary to measure B in the unit mev = millions of electron volts. The relation between mass on the O^{16} scale and this energy unit is 1 millimass unit = 0.931 mev.

In Fig. 1.1 the *binding energy per particle* B/A is plotted as a function of A for stable nuclei. The binding energy per particle is roughly constant for medium and heavy nuclei at about the value of 8 mev per nucleon. This behavior contrasts sharply with that of the electron binding energy per electron in atoms, which increases (irregularly) with the

1.2 · NUCLEAR MASSES AND BINDING ENERGIES

number of electrons in the neutral atom. With respect to the binding energy, the nucleus is similar to a solid or a liquid for which the heat needed for vaporization is proportional to the mass of the substance.

It is useful to consider, in addition to the binding energy B, or the *binding fraction* B/A, the separation energies of nucleons (neutrons (n), protons (p)) or nuclear aggregates (deuterons, H_1^2, α particles, He_2^4 etc.) from a nucleus. The separation energy of a particle b (nucleon or aggregate) is defined as the smallest energy required for the separation of a nucleus into two parts, one of which is the particle b. If the residual

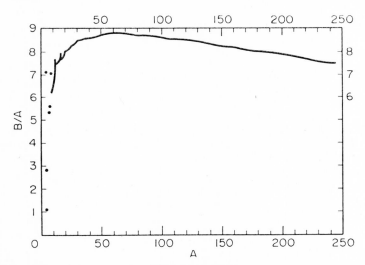

Figure 1.1. Binding Energies of Stable Nuclei. The binding energy, in mev per nucleon, is plotted against the number A of nucleons.

nucleus after the separation is designated by the index r, the separation energy is given by:

$$S_b = [M_b + M_r - M(N, Z)]c^2.$$

The separation energies characterize the relative stability of a nucleus with respect to transformations leading to the emission of a nucleon or nuclear aggregate.

The separation energies of neutrons S_n and of protons S_p are of the order of the binding fraction for stable nuclei. The separation energy S_α of an α particle (He_2^4) is, for intermediate A nuclei, of the order of 5 mev. For the unstable α-radioactive nuclei the S_α is negative. (See Section 1.3).

As already mentioned, the smooth variation of B/A with A suggests that the nucleus has properties similar to those of a liquid or solid. The fluctuations in S_b, however, indicate the inadequacy of such pictures. A close examination of the separation energies S_p and S_n of stable nuclei as functions of Z and N reveals discontinuities which are similar to those which occur in the electron separation energy (ionization potential) of stable atoms. Thus the separation energy of a neutron from Si^{28} is 17.2 mev, that of a proton from N^{13} is 1.95 mev. However, these are extreme cases, and on the whole the fluctuations in the separation energies S_p and S_n are less marked than those in the ionization energies of atoms. (See Chapters 2, 7). The largest separation energies S_p and S_n (20 mev) are those for the α particle, He_2^4; as a result, He_2^4 has exceptional stability.

1.3. Types of Nuclear Instability. Spontaneous and Induced Transformations

Certain nuclei among those which occur naturally undergo spontaneous transformations. Thus the naturally radioactive substances with $Z > 82$ (Pb) exhibit transformations with emission of α particles, of electrons, and of electro-magnetic radiation (γ rays). The processes are often referred to as α-, β-, and γ-decay processes, respectively. Not only the heavy natural nuclei undergo spontaneous transformation. For example, K^{40} also emits electrons; this nucleus also transforms by the absorption of an extra-nuclear electron—the K-capture process (since a K electron of the atom is absorbed).

Unstable states of nuclei can also be formed in nuclear reactions. Nuclei can be formed which subsequently transform by emission of positive or negative electrons or by capture of an orbital electron. Nuclei may be excited in collision processes with subsequent emission of γ rays. It is also possible to form nuclear states (generally of very short lifetime—see below), which transform by emission of neutrons, protons, or α particles.

All the transformations mentioned above may be characterized by a transformation or decay probability—the probability per unit time of transition between initial and final states of the system. The reciprocal of this probability measures the *lifetime* of the initial state. The lifetime is defined as the time required for the population of the initial state to fall to e^{-1} (e is the base of natural logarithms) of its initial value. The emission probability λ is often expressed in terms of the level width instead of in units of reciprocal time; the width Γ, which may be considered as an uncertainty in the energy of the unstable state, is related to the average or mean decay time $\bar{\tau} = \lambda^{-1}$ through Heisenberg's relation, $\lambda = \Gamma/\hbar$. The relation between Γ (in electron volts) and λ (in $\sec^{-1}$) is $\Gamma = 0.65 \cdot 10^{-15} \lambda$. The "half life" is the time required for the transformation of one half of the original material; it is $\tau_{1/2} = \bar{\tau} \ln 2 = 0.693 \bar{\tau}$. The general characteristics of the various types of nuclear instability are summarized below.

a. **Natural α-radioactivity; Fission.** The range of α-decay lifetimes is very broad. The lifetime is a sensitive function of the energy of the emitted α. The lifetime of Th^{232} is $\sim 10^{10}$ years for an α energy of 4 mev; the corresponding numbers for Po^{212} are 3×10^{-7} sec, and 9 mev.

The characteristics of the relation between energy and lifetime may be understood in terms of the concept of barrier penetration. Consider the potential energy as a function of the distance of the α particle from the residual nucleus which remains after transmutation. For large distances the potential energy is purely electrostatic. With decreasing r the potential energy increases as r^{-1}. When the distance is of the order of, or smaller than, the nuclear radius the interaction is no longer purely electrostatic. Strictly speaking, the concept of the potential energy of the particle loses significance for such small distances since the α particle does not maintain its identity and structure within the nucleus. We shall, however, consider the α particle as moving in some effective potential even inside the residual nucleus. At some separation the attractive nuclear forces overcome the electrostatic forces, leading to an effective potential like that given in Fig. 1.2. In order to be released from the nucleus, the α particle must penetrate the electrostatic barrier; that is, it must pass through the region where the potential energy is higher than the total energy of the α particle. On the basis of the theory of barrier penetration, the relation between the energy of the α particle and the lifetime of the unstable nucleus is readily understood. (Section 9.5.)

The height and breadth of the assumed electrostatic barrier depend

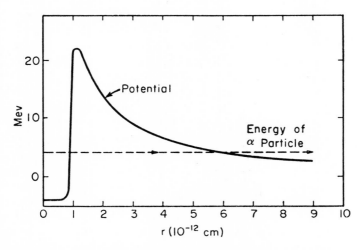

Figure 1.2. Potential of an α particle as function of the distance from the center of a nucleus (schematic). The potential energy is given in mev; the energy of the α particle, as indicated, is a typical disintegration energy.

critically on the value chosen for the nuclear radius, i.e. the radius at which the electrostatic forces are overcome by internal nuclear forces. This in turn strongly affects the theoretical results concerning the probability of barrier penetration. The comparison of theoretical predictions with observed lifetimes thus provides a means for determining the radii of α-radioactive nuclei.

It may be of interest to note that the separation energy of an α particle is negative for many heavy elements which do not exhibit α-activity. The reason is that the barrier is so effective at the low α-emission energies available, that the transformation probabilities are too small for the transformations to be observed.

The binding fraction curve indicates the possibility of another type of instability for the heavy nuclei. The masses of these nuclei are larger than the masses of the nuclei obtained by roughly splitting them in two. Thus "fission" of the heavy nuclei, i.e. splitting into two approximately equal parts, becomes possible with the release of considerable energy ($\sim$200 mev). However, the barrier which impedes the process is of such height that the probability of transformation by splitting is very small. For the heaviest nuclei an excitation of the nucleus by a few mev takes the system over the barrier and fission follows within a very small fraction of a second.

b. γ-radiation; Particle Emission. If a stable nucleus is excited but has insufficient energy for the emission of a particle, it will return to its normal state by emission of electromagnetic radiation. Lifetimes for such processes are extremely sensitive to spin changes between initial and final states (the energy dependence, which favors high-energy changes, is less important). The lifetimes for γ-emission range from 10^{-17} to 10^{-10} seconds if the spin (J) change in the transition is small; i.e. $\Delta J \leq 2$. For low-energy transitions with large spin changes the lifetime can become abnormally long. For $\Delta J = 4$ and energy change $\sim$0.1 mev the lifetime can be several years. If the first excited level of a nucleus has a long life, the excited state of the nucleus is called an isomeric state.

In practice, the distinction between states which are unstable only with respect to the emission of γ rays and those which are unstable also with respect to particle emission is very important. Theoretically, the two processes take place side by side, each with its characteristic probability. For *light elements* the particle emission, if it is at all possible energetically, is usually so much more probable than the emission of γ rays that the latter can be neglected altogether. This is also true for neutron emission in heavier elements except when the neutron emission is only "barely possible," i.e. if the energy of the emitted neutrons is less than a few kilovolts. However, the emission of "low-energy" charged particles from *heavy nuclei* is so much impeded by the potential barrier that, as a rule, γ ray emission is the faster process.

Just as the total decay probability is equal to the sum of the decay probabilities of the various processes (γ emission, emission of various particles) by which the nuclear state in question can decay, the total width of the state is the sum of the "partial widths" corresponding to the same processes. (Chapter 9.)

Examples of particle instability other than those of natural radioactive elements are: the disintegrations of Be^8 into two α particles, and the disintegrations of He^5 and Li^5 into an α particle and a neutron and proton, respectively. In addition, practically every nucleus has excited states with sufficient excitation energy to permit the emission of some of its constituents.

c. β-decay. Perhaps the most interesting forms of nuclear instability, both theoretically and experimentally, are those involving the emission of an electron or positron (β-decay) or the capture of an orbital electron (K-capture). Lifetimes of these processes range from around 10^{-2} sec to more than 10^{11} years. The emitted electrons do not carry the whole energy difference between initial and final states (as do α particles). Instead, they have a characteristic energy distribution (the Fermi distribution); the upper limit of this energy distribution is, to within the errors of observation, equal to the energy made available in the transition.

Since electrons are not fundamental nuclear constituents, they must be created in the β-transformation. The process is analogous to the creation of photons in radiative transitions. The conservation of energy, momentum, and angular momentum in the β-process, and the distribution in energy of the emitted electrons can be understood only if it is assumed that, in addition to the electron (or positron), a second particle (called the neutrino and designated by the symbol ν) is emitted. To account for the charge, energy, and angular momentum conversation in β-transitions, it must be further assumed that the neutrino is uncharged, has a mass less than a thousandth of the electron mass, and possesses an intrinsic spin angular momentum $\frac{1}{2}\hbar$. These assumptions originally put forward by Pauli enabled Fermi to explain the characteristic energy distribution of the β electrons. (A more detailed account of this theory is given in Chapter 11.) The interaction of neutrinos with matter is so weak that they penetrate all matter practically unhindered. Direct evidence for the existence of the neutrino has been obtained only very recently by observing its interaction with nuclei different from the emitting nucleus.

In the β-decay processes the number of nucleons, A, is the same in the initial and final nuclear states. The transition involves the transformation of a neutron (or proton) in the original nucleus into a proton (or neutron) in the final nucleus. Thus the β-transitions may be indicated by:

$$(A, T_\zeta) \to (A, T_\zeta \pm 1) + e^{\mp} + \nu) \text{ (positive or negative electron emission)}$$

$$(A, T_\zeta) + e^- \to (A, T_\zeta + 1) + \nu \text{ (capture of the orbital electron),}$$

where the symbol (A, T_ζ) designates the initial nuclear species. The energy available in the negative electron emission or capture process is equal to the difference in *atomic masses* of the initial and final nuclei. (For positron emission this mass difference must exceed $2mc^2$ (1.1×10^{-3} mass units) if the process is to be energetically possible). Thus two neighboring isobars ($\Delta T_\zeta = 1$) cannot both be stable with respect to β-decay. For the case of $A = 1$, for example, the neutron mass is greater than that of the hydrogen atom; the neutron decays with emission of negative electrons ($\tau_{1/2} = 13$ min). The theory of β-decay shows that the lifetime depends upon the available energy E, roughly as E^{-k}, where k is approximately 5. Thus the decay time for β-disintegration depends on the decay energy much less sensitively than that of α-disintegration, so that low-energy β-processes are not in general too strongly inhibited to be observed. However, the lifetime depends sensitively on spin and parity changes between initial and final nuclear states. For large spin change and low energy the lifetime may well become so long that the nucleus appears to be stable. Thus, one of the two $A = 50$ isobars must be unstable, yet neither exhibits observable β-activity. The same applies for the pair Cd^{113}, In^{113}. An interesting case of an apparently stable isobar is Lu^{176} which has a half-life of about 4×10^{10} years. It was one of the odd-odd nuclei, the radioactivity of which was difficult to discover. In^{115} and Re^{187} were also found to be radioactive with half-lives of 6×10^{14} and about 10^{10} years, respectively.

In contrast to the scarcity of isobaric pairs with $\Delta T_\zeta = 1$, there are many examples of pairs of isobars with T_ζ differing by two units. The higher mass nucleus of such a pair may in principle decay to its partner by a double β-process in which two electrons (positrons) are emitted (or by a double K-capture). The theory of this process depends on more detailed assumptions concerning the nature of the neutrino than the theory of ordinary β-decay. Two theories have been put forward. In the first the double decay gives rise to two neutrinos as well as two electrons, and the electrons have a continuous distribution of energy up to the limit permitted by the mass difference between the isobars. The lifetimes for this process are extremely long—of the order of 10^{20} years for an available energy difference of 5 mev. In the second theory

1.3 · NUCLEAR REACTIONS AND DISINTEGRATIONS

the double β-process may occur without neutrino emission because the neutrino emitted in the first β-decay is reabsorbed in the second. If this process occurs—it implies a modified theory of the individual neutrino—the sum of the energies of the electrons is equal to the whole disintegration energy, and therefore is a constant. The lifetimes may be as short as 10^{14} years for such processes under favorable circumstances. Even this lifetime is still sufficiently long to account for the existence of ostensibly stable isobaric pairs.

Search for the double β-decay process (which should be just within the limits of observability with present techniques if the second theory is correct) has been made repeatedly. Results seem to speak against the existence of the second kind of process, and therefore in favor of the standard theory of the neutrino as an ordinary particle with zero rest mass and spin 1/2.

CHAPTER 2

Systematics of Stable Nuclei. Details of Binding Energy Surfaces

Study of the stable nuclei reveals many interesting and remarkable regularities. For $A < 36$ the stable nuclei have almost equal numbers of protons and neutrons. The T_{ζ} values for all A in this region are ≤ 1. With increasing A the T_{ζ} values increase regularly. For heavy nuclei ($A \sim 200$) T_{ζ} is about 25. Fig. 2.1, the so-called Segré chart, gives all nuclides occurring in nature. Most of these nuclides are stable. Some of them exhibit instability by α- or β-radioactivity; some of them are probably unstable, but owing to their long lifetime their radioactivity has not been observed. All nuclei with $A > 209$ are radioactive.

There are many more stable nuclei with even A than with odd A. Moreover, the nuclei with even A almost invariably have even Z and even N. Indeed, only four odd Z, odd N stable nuclei are known, and all of these have $T_{\zeta} = 0$. The odd-odd nuclei are H_1^2, Li_3^6, B_5^{10}, N_7^{14}. (Two other cases V_{23}^{50} and Ta_{73}^{180} are believed to be β-unstable with unobservably long lives). The fact that even numbers of neutrons or protons lead to more stable structures can be seen from a counting of the numbers of stable isotopes for even or odd Z, or the number of stable isotones for even or odd N. For odd Z there are never more than two stable isotopes; for even Z the number ranges as high as ten. There are no stable isotopes for $Z = 43$ or 61, while there is only one even Z nucleus, Be_4^9, for which there is only one stable isotope. Above $A = 14$ there are only two odd N values for which a pair of stable isotones occur, ($N = 55$, 65 and $N = 27$ and 107 if V_{23}^{50} and Ta_{73}^{180} are stable). Otherwise the number of stable isotones for odd N is one or zero (the number is zero for $N = $ 19, 21, 35, 39, 45, 61, 89, 115, 123). There are stable nuclei for all even N; for $N = 82$ the number of isotones is 6 (for $N = 50$, it is 5). For any given A the number of isobars is small. There is one stable isobar for each A between $A = 1$ and 36 except for $A = 5$, 8 for which there are none.

The number of isobars with odd A is generally one. In those cases where two are observed, it is believed that one of the isobars is β-unstable (the ΔT_{ζ} for these isobaric pairs is 1). About fifty isobaric pairs are known for even A; the T_{ζ} values of the pairs differ by two. For $A = 96$, 124, 130, and 136 there are three isobars. In these cases also the difference in T_{ζ} between neighboring stable isobars is two.

The even-odd relationships indicated above and the systematics of

2 · PROPERTIES OF STABLE NUCLEI

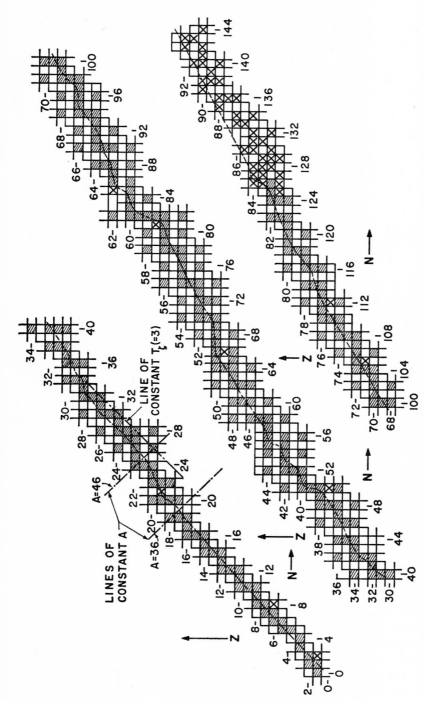

Figure 2.1. Chart of Natural Nuclei. Each square corresponds to a charge number Z and a neutron number N. The squares which correspond to stable nuclei are shaded, those of natural radioactive nuclei bear an $\times$.

isobars can be understood on the following basis. It must be assumed that binding energies of nuclei near the stable region can be represented not by one surface $B(N, Z)$ but by three close surfaces: one for the even-even, one for even-odd and odd-even, and one for odd-odd nuclei. Note that in the β-decay process the even-odd nuclei transform into odd-even nuclei, while odd-odd nuclei transform into even-even ones, or conversely. Existence of stable isobaric pairs of even-even nuclei and the absence of stable odd-odd nuclei shows that the mass curves (as a function of T_ζ for constant A) are roughly as shown in Fig. 2.2.

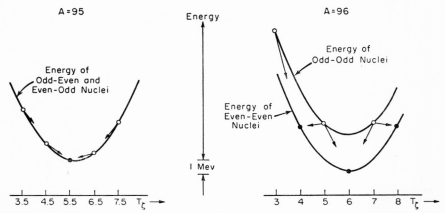

Figure 2.2. Masses of Isobaric Nuclei. The diagram on the left illustrates the energy (mass) of isobars with odd mass number A, as function of the difference between neutron and proton numbers; $T_\zeta = (N - Z)/2$. The diagram on the right applies for even A. In this case there are two lines: one for nuclei with even N and even Z, the other for nuclei with odd N and odd Z. The open circles represent unstable nuclei, the full circles stable nuclei. The energy of a stable nucleus is lower than that of either adjacent isobar. There is only one stable isobar in the case of odd A (in the diagram the isobar with $T_\zeta = 11/2$; there is no stable nucleus with odd N and odd Z; there are, under the conditions of the diagram on the right, three stable isobars with even N and even Z.

For the case illustrated all odd-odd nuclei are unstable and three even-even isobars will be stable. This picture also explains the fact that the T_ζ difference between stable isobars is almost always two. From the energies of the β-transitions from odd-odd to even-even nuclei, the displacement between the even-even and odd-odd mass curves may be measured; it is about 68 mev/$A^{3/4}$. Reaction experiments locate the mass curve of the even-odd nuclei about midway between the even-even and odd-odd curves. The fact that the odd A nuclei exhibit only one stable isobar for each A indicates that the mass curve is the same for the odd N, even Z, and even N, odd Z nuclei.

A number of fairly good semi-empirical binding energy or mass functions have been constructed which give a good approximation to

2 · PROPERTIES OF STABLE NUCLEI

the true masses over the wide range of both stable and unstable nuclei in terms of relatively few empirical constants. Such an approximation is provided by Weizsäcker's formula:

$$B(Z, N) = U_v A - U_c Z(Z - 1)A^{-1/3}$$
$$- U_s A^{2/3} - U_t T_\zeta^2/A \begin{cases} + \delta/A^{3/4} & \text{even-even} \\ + 0 & \text{even-odd} \\ & \text{or odd-even} \\ - \delta/A^{3/4} & \text{odd-odd} \end{cases} \quad (2.1)$$

The first term provides a "volume" effect, i.e. a binding energy proportional to the number of nucleons. The second term expresses the fact that the binding energy is diminished by the electrostatic repulsion between the protons. There are $\tfrac{1}{2}Z(Z - 1)$ interactions between the charges which are taken to be uniformly distributed over the volume of the nucleus. The $A^{-1/3}$ factor expresses the dependence of this energy contribution on the radius of the nucleus, which is assumed to be proportional to $A^{1/3}$. The third term expresses the diminution of the volume energy contribution due to surface effects. Terms similar to the first three occur also in the expression for the energy of a charged liquid drop. The fourth term takes into account the effects of symmetry properties of the nuclear states and how these are modified when the number of nucleons is kept constant and T_ζ, i.e. the neutron excess, is varied. The last term also represents a symmetry effect. Constants which provide a fit with the binding energy data are:

$$U_v = 14.0 \quad \text{mev} \quad U_c = 0.61 \quad \text{mev} \quad U_s = 14.0 \quad \text{mev}$$
$$U_t = 84.2 \quad \text{mev} \quad \delta \sim 34 \quad \text{mev.} \quad (2.1a)$$

The existence of three binding energy surfaces provides a qualitative understanding of the distribution of isobars. It also explains why the separation energy of a neutron $S_n = B(Z, N) - B(Z, N - 1)$ is greater for even N than for odd N, and why the separation energy of a proton $S_p = B(Z, N) - B(Z - 1, N)$ is greater for even than for odd Z. For even N, the S_n contains the $\delta/A^{3/4}$ term with positive sign; for odd N with negative sign, and a similar remark applies for the separation energy of the proton. However, as long as we compare S_n or S_p only for nuclei with even N and even Z (or, more generally, as long as we do not change the even or odd nature of Z and N), both S_n and S_p as calculated from (4.1) will be smooth functions of Z and N. Measurements of the separation energies show that this is true in general. However, there are a number of significant exceptions which are connected with the so-called *magic numbers* 2, 8, 20, 50, 82, 126, and to a lesser degree, some others. When Z passes these numbers, S_p undergoes a drop of about 2 mev. Thus, the separation energy of a proton from Sn_{50} ranges from $8\frac{1}{2}$ to $10\frac{1}{2}$ mev as A increases from 114 to 120. For Te_{52} these numbers range from 8 to 9 mev as A increases from 122 to 126. The increase of S_p with increasing A is due to the increasing number of neutrons which bind a constant number of protons with increasing strength. In spite of this increase the S_p is higher for the "magic" Sn_{50} nuclei than for the non-magic Te_{52}. The comparison of the In_{49} and Sb_{51} nuclei is less convincing: for the former S_p ranges from 6 to $7\frac{1}{2}$ mev. For the latter, which has a single proton in excess of the magic 50, S_p ranges from $5\frac{1}{2}$ to 7 mev. Note the generally smaller value of S_p for odd than for even Z: the even-odd difference is larger than the magic — non-magic difference. The situation at the other magic numbers is perhaps even more striking, and a similar phenomenon is noted with respect to S_n when N passes through one of the magic numbers. The most marked discontinuity occurs at $N = 2$ and $Z = 2$ where both S_n and S_p drop from 20 mev to about 7 mev. The explanation of these discontinuities goes beyond the scope of global formulae such as (2.1) which do not take into account properties of individual nucleon orbits. An explanation of the properties of the magic numbers is given in Chapter 7 by assuming that the nucleons are arranged in shells similar to the K, L, M, etc. shells of electrons in the outer regions of the atoms.

The particular stability of the magic numbers manifests itself also in the Segré chart. The number of isotones for magic N or isotopes for magic Z is larger than the number of isotones or isotopes for adjoining even values of N and Z. The difference is even larger if comparison is made with adjoining odd values of N and Z, but this is a consequence of the $\delta/A^{3/4}$ terms in (2.1). The preferred stability of the magic N and Z numbers, as compared with adjoining even N and Z numbers, on the other hand, does not follow from (2.1) but only from the particular "magic" stability of these neutron and proton numbers.

CHAPTER 3

Properties of Nuclear States; Ground States

3.1. Spins and Moments

Although much painstaking work has gone into investigation of excited states of nuclei, our knowledge and understanding of these states is still far from complete. The spectroscope can be employed to obtain, quite rapidly, a tremendous wealth of data on the excited states of atoms. Unfortunately, no instrument of corresponding power exists for nuclei.

Over the past several years a considerable effort has been expended in the measurement of the properties of nuclear ground states for which special techniques are available. In addition to a definite energy, a stationary or quasi stationary state also has a definite angular momentum characterized by the quantum number J, (the square of the angular momentum is $J(J + 1)\hbar^2$; the number J is referred to as the *spin*), and a parity. These properties are common to all stationary states of free systems. The spin J may be measured by a number of techniques. A few of the regularities of the spins of nuclei follow. The spin of all even-even nuclei is zero. No exception to this rule is known. The even-odd and odd-even nuclei always have half integral spin, while odd-odd nuclei have integer spin values. This strongly supports the assumption that nuclei are composed of neutrons and protons. The spins which follow from the earlier proton-electron composition theories are inconsistent with this rule. The spins of the stable nuclear ground states are in general rather small. The largest measured value for an odd nucleus is 9/2.

Measurement of the electric quadrupole and magnetic dipole moments has greatly advanced our knowledge of nuclear structure. The magnetic 2^l poles vanish for even l (for free systems), while the electric 2^l poles vanish for odd l. Hence, the magnetic dipole moment and the electric quadrupole moment are the lowest measurable moments. Apart from a few magnetic octupoles which have been obtained by a careful analysis of hyperfine structure patterns, no higher moments have been measured. It follows from the general properties of the $J = 0$, $J = \frac{1}{2}$ states of a free system that the magnetic dipole moment vanishes for $J = 0$ while the quadrupole moment is zero for $J = 0$, or $\frac{1}{2}$.

The magnetic dipole moment μ of a nucleus will affect the energy of a nuclear state in a uniform magnetic field. The energy change of the nucleus when its spin is aligned in the direction of the field H is $-\mu H$.

3 · PROPERTIES OF THE NORMAL STATE

A measurement of the energy change determines the dipole moment μ. For a single nucleon moving in a central field the magnetic dipole moment can be calculated if the spin and angular momentum of the nucleon are known. For a single particle (proton or neutron) of definite orbital angular momentum and definite total angular momentum—specified by the quantum numbers l and $j = l \pm \frac{1}{2}$—the magnetic moment in units of the nuclear magneton $e\hbar/2M_p c$ is given in Figs. 3.1a, b. The curves in these figures are known as the Schmidt lines.

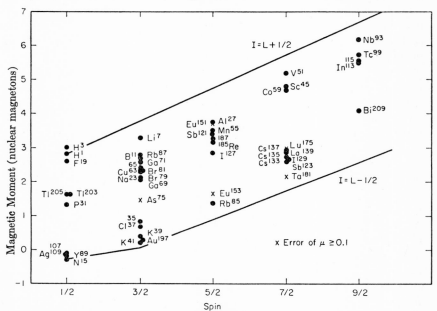

Figure 3.1a. Magnetic Moments of Nuclei with even N and odd Z. The magnetic moments, in units of $e\hbar/2M_p c$, are plotted against the spin. The two full lines are the Schmidt lines; most magnetic moments are between these.

The points on Fig. 3.1 indicate the measured moments. The observed magnetic moments of nuclei with odd Z and even N lie between the proton Schmidt lines (except for H^3), while the nuclei with odd N even Z lie between the neutron Schmidt lines (except for He^3 and C^{13}). The magnetic moments of even A nuclei are generally zero since, except for the few stable odd-odd nuclei, the spins of even A nuclei are zero.

The energy of a system the charge distribution of which is axially symmetric will depend on the gradient of the electric field parallel to the symmetry axis. The part of the energy which depends on the relative orientation of the field and system is:

$$E = -\frac{1}{4}\left(\frac{\partial \mathcal{E}_z}{\partial z}\right)_0 eq, \qquad (3.1)$$

(terms in the higher field derivatives associated with higher order poles have been dropped), where $\mathcal{E}$ is the electric field and q is the quadrupole moment of the axially symmetric distribution divided by e. The zero indicates that the value of the gradient is to be taken at the center of the nucleus. In terms of the density distribution $\rho(\mathbf{r})$ of the protons,

$$q = \int \rho(\mathbf{r})(3z^2 - r^2) \, dV, \qquad (3.2)$$

where the integration is carried out over the whole nucleus. The q for the nuclear state the spin of which is oriented along the z axis is called the nuclear quadrupole moment. Values for the observed quadrupole moments are given in Fig. 3.2. Except for nuclei with very large quadru-

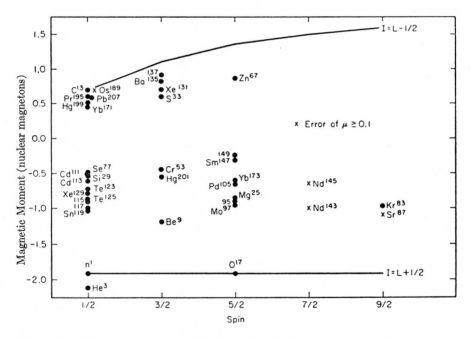

Figure 3.1b. Magnetic Moments of Nuclei with odd N and even Z. See description of Figure 3.1a.

pole moments, distortion of the nucleus from spherical symmetry as indicated by the quadrupole moment is rather small. The "normal" quadrupole moments are as large as that of a uniformly charged ellipsoid of revolution with a difference of major and minor axes amounting to about one or two percent of the nuclear radius. For the anomalous quadrupole moments this difference rises to ten or fifteen percent and even more.

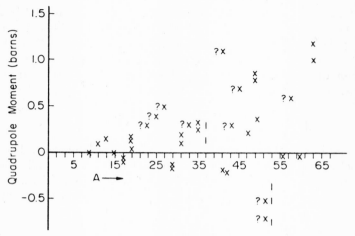

Figure 3.2. Quadrupole Moments of Nuclei with odd mass numbers A, in barns (10^{-24} cm²).

3.2. The Size of the Nuclei

A variety of methods can be used to explore nuclear radii and are found to provide consistent results. The nuclear size can be estimated (1) from the scattering cross-sections of nuclei for fast neutrons (neutron wave length $\lambda \ll$ nuclear radius), (2) from the energy-lifetime relationship in α-decay (Section 1.3), (3) from the difference of the binding energies of "mirror nuclei," (4) the energy levels of mesic atoms, and (5) from high-energy electron diffraction experiments.

The first method, neutron scattering, depends on the assumption that the total cross-section for high-energy neutrons is $2\pi R^2$, where R is the nuclear radius. In method (2) the theory of the penetration of the electrostatic barrier shows that the disintegration probability depends on the "radius" at which specifically nuclear forces overcome the electrostatic forces. The value of this radius which gives agreement with experiment is the nuclear radius. Methods (1) and (2) give only semiquantitative measures of the nuclear size. Method (3) requires a slightly more extended explanation. A pair of nuclei are called mirror nuclei if they have the same A, and the T_ζ of one is the negative of the T_ζ of the other. The composition of the second nucleus is obtained by changing all nucleons of the first from protons to neutrons and from neutrons to protons. On the assumption that the only difference between the proton-proton and neutron-neutron interactions is given by the electrostatic repulsion between the former, the kinetic energy and the potential of the nuclear forces will be the same in both nuclei. Their masses will differ only for two reasons: the greater number of neutrons in the positive T_ζ nucleus increases its mass because the neutron is heavier than the proton; on the other hand the electrostatic potential is greater in the negative T_ζ nucleus because this contains more protons. Except for the H^3-He^3 pair this latter effect is larger than the former. It can be estimated on the assumption that the electric charge of the nucleus is uniformly spread over a sphere of radius R. (A small correction for exchange effects must be introduced.) Hence the last and most important contribution to the mass difference of mirror nuclei depends on the nuclear radius; it is inversely proportional thereto. Since the mass difference in question can be obtained experimentally (for instance from the energy of the β-decay of one into the other), one obtains a measure for the nuclear radius. The radii obtained can be approximated by:

$$R = A^{1/3} r_0 \qquad r_0 = 1.37 \times 10^{-13} \text{ cm.} \tag{3.3}$$

This gives for the root mean square distance of the nucleons from their center of mass $(\tfrac{3}{5})^{1/2} R = 1.05 \times 10^{-13} A^{1/3}$ cm. Since R is proportional to the cube root of A, the densities of all nuclei are approximately the same.

When μ mesons are slowed down in a material, they may live long enough to be captured in orbits which are similar to the electronic orbits in atoms. The energy differences between these orbits, on which the fourth method of measuring nuclear radii is based, can be obtained by measuring the energy of the gamma rays which are emitted in the course of the transitions between these orbits. The "radius" of the innermost *electronic* orbit of an atom is $\hbar^2/Ze^2 m$. This is, even for the heaviest nuclei (largest Z), about a hundred times larger than the nuclear radius. As a result the electrostatic potential acting on the electron is very nearly that of a point charge, and the energy values of the electrons are not affected appreciably by the finite size of the nucleus. The picture is markedly different however for the orbits of mesons. The formula $\hbar^2/Ze^2 m$ again gives the "radius" of the lowest orbit, but m is in this case the mass of the meson rather than that of the electron. Since the mass of the μ meson is about 207 times greater than the mass of the electron, the radii of the smallest mesonic orbits around heavy nuclei are quite comparable with the radii of these nuclei. As a result, the electrostatic potential acting on a meson in its smallest orbit differs greatly from the potential of a point charge—it is more nearly the potential of a reasonably large, uniformly charged sphere. The energy levels therefore show a considerable or even very large displacement from the positions which they would have if the electrostatic potential of the nucleus were that of a point charge. From the size of the displacement the radius of the charge distribution can be calculated. This method yields the value $1.2 \times 10^{-13} A^{1/3}$ cm for the nuclear radius, if the nucleus is assumed to be spherical and to have a uniform charge distribution. The corresponding mean square radius of the protons can be calculated to be:

$$\overline{r^2} = \tfrac{3}{5}[1.2 \times 10^{-1/3} A^{1/3}]^2 = 0.85 \times 10^{-26} A^{2/3} \text{ cm}^2. \quad (3.4)$$

This is considerably less than the values given by the preceding methods. It should be noted, however, that this method is accurate only for heavy nuclei (large Z). For light nuclei even the mesonic orbits lie outside the nucleus. As a result the electrostatic potential at the orbit is so close to that of a point charge that the energy shift of the orbit is too small to be measured accurately.

The interpretation of the measurements of the energy levels of μ mesons presupposes that the interaction between nucleons and μ mesons is of purely electrostatic nature. If other forces also acted between μ mesons and nuclei they would also change the mesonic orbits and energy levels, and the relation between energy levels and nuclear radii as developed above then would require modification. There is, at present, no reason to believe that other than electromagnetic forces play a significant role in the μ meson-nucleon interaction. The same is true of the interaction between electrons and nucleons. This makes it possible to investigate the charge distribution in nuclei by electron scattering experiments. Naturally, as long as the wave-length of the electrons is very large as compared with the size of the nucleus, the scattering will closely resemble scattering by a point charge. However, if the wave-length of the electrons is comparable to the diameter of the nucleus, wavelets scattered from different parts of the nucleus will interfere—in some directions constructively, in others, destructively. Under such conditions—which are obtained if the electron energy is of the order of 100 mev—the scattering will differ considerably from that of a point charge, i.e. from Rutherford scattering. By analyzing the angular distribution of the scattered electrons one can, in principle, obtain not only the size of the nucleus (i.e. a number such as the mean square distance of the protons from their center of mass), but the whole charge distribution. This is the basis of the fifth method of measuring nuclear radii.

The best high-energy electron scattering experiments have been made by Hofstadter and his collaborators. Their interpretation by Yennie, Schiff, and their collaborators does not yet give the functional dependence of the proton density as a function of the distance from the center. However, it does allow the determination of two constants characterizing the proton distribution, in contrast to the single constant given by all other measurements. These two constants can be interpreted as the mean square distance of the protons from their center of mass, and the diffuseness of the nuclear surface. This last quantity $2z$ is, crudely, the distance in which the proton density drops from $\tfrac{3}{4}$ of

3.2 · SIZE OF NUCLEI

its maximum value to $\frac{1}{4}$ of this value. A few values of these quantities follow (in units of 10^{-13} cm):

A	12	40	51	59	115	122	197	209	
$[\overline{r^2}/A^{2/3}]^{1/2}$	1.05	1.02	0.97	0.98	0.92	0.93	0.91	0.93	$\times 10^{-13}$ cm
$2z$		1.1	1.0	1.1	1.0	1.1	1.1	1.2	$\times 10^{-13}$ cm

One notices that $\overline{r^2}$ increases somewhat less fast than $A^{2/3}$, i.e. that the average density increases somewhat with increasing size of the nucleus. This increase appears to be due to the decreasing role of the nuclear surface at which the density is, naturally, smaller than in the inside of the nucleus. The mean square radius, as given by the electron diffraction measurements, is consistent with the radii given by other methods.

The general agreement of the radii obtained by the different methods serves also to substantiate theories of barrier penetration and the theory of the equivalence of neutron-neutron and proton-proton interactions.

CHAPTER 4

Survey of Nuclear Reactions

4.1. Types of Reaction, Cross-Sections, Excitation Functions

In a broad class of nuclear experiments, studies are made of the collisions of neutrons (n), protons (p), deuterons (d), and α particles with various other nuclides. When the bombarded nucleus is transformed in any way in the collision the process is called a nuclear reaction; otherwise one speaks of elastic scattering. If the species X is bombarded by a particle a to form a nucleus Y and fragment b, the reaction is described by $a + X \rightarrow Y + b$, or more compactly by $X(a, b)Y$. The probabilities of various reactions or scattering processes may be characterized by *cross-sections*. The cross-section σ of a process is defined as the probability that the process occurs if the incident beam consists of a single particle and the target contains one nucleus per unit area. The number of processes is proportional to the number of particles in the incident beam and to the number of target nuclei per unit area of the target. For a given pair of colliding nuclei there are, in general, a number of possible reaction products. The total cross-section for the process is the sum of the cross-sections for the individual processes.

A convenient unit for nuclear cross-sections is the *Fermi* or *barn*, an area of 10^{-24} cm^2, which is approximately equal to the cross-sectional area of medium A nuclei.

One is often concerned with the probability of the production of a reaction product within an element of solid angle $d\Omega$ about a given direction; if $d\sigma$ is the cross-section for this process, then $d\sigma/d\Omega$ is called the differential cross-section (it is often designated simply as $d\sigma$). In a moving coordinate frame with respect to which the center of mass of the collision system is at rest, the two products of the reaction move off in opposite directions. Since the collision experiment (if the incident and target nuclei are not polarized) is characterized by axial symmetry about the direction of the beam, the differential cross-section is a function only of the angle θ between the line of motion of the incident particle and the line of motion of the product particle in the center of mass coordinate frame. The differential cross-section $d\sigma(\theta)$ as a function of θ gives the angular distribution of the reaction products. The differential cross-section is usually measured for the lighter of the two reaction products in the laboratory frame of reference. Fig. 4.1 illustrates the connection between the velocity diagrams of a reaction in the laboratory frame of reference and in the center of mass coordinate system.

4.1 · CROSS SECTIONS

Coordinate System	Time	Velocity Relations	Velocity Diagrams	Energy of Target Nucleus	Total Kinetic Energy
At Rest	Before Collision			0	$\frac{1}{2} M_a v_i^2$
Center of Mass	Before Collision	$M_a v_i / (M_a + M_x)$ is Subtracted from Both Velocities	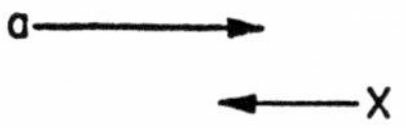	$\frac{\frac{1}{2} M_x M_a^2 v_i^2}{(M_a + M_x)^2}$	$\frac{\frac{1}{2} M_a M_x v_i^2}{M_a + M_x}$
Center of Mass	After Collision	Only Direction of Velocities Changes	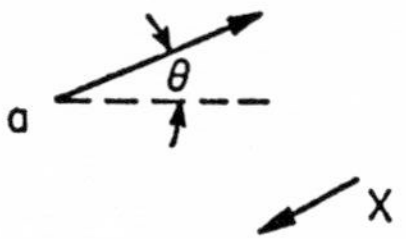	$\frac{\frac{1}{2} M_x M_a^2 v_i^2}{(M_a + M_x)^2}$	$\frac{\frac{1}{2} M_a M_x v_i^2}{M_a + M_x}$
At Rest	After Collision	$M_a v_i / (M_a + M_x)$ is Added to Both Velocities		$\frac{2 M_x M_a^2 v_i^2}{(M_a + M_x)^2} \sin^2 \frac{1}{2}\vartheta$	$\frac{1}{2} M_a v_i^2$

Figure 4.1. Reduction to the Center of Mass Coordinate System in Binary Collisions.

The difference between the internal energies of the reacting pairs a, X and the final products Y, b is called the Q of the reaction. If Q is positive (exothermic reaction), the reaction is energetically possible even in the limit of zero bombarding energy; if Q is negative (endothermic reaction), the reaction can occur only if the kinetic energy of the colliding particles in the center of mass coordinate system exceeds the energy difference between the Y, b and the X, a pairs. This energy difference is called the *threshold* of the reaction. The dependence of a cross-section $\sigma(E)$ on the energy E is often referred to as the *excitation function*. Energy in this connection means the kinetic energy of both particles in the center of mass coordinate system. This is smaller than the energy E_{lab} of the single bombarding particle: one sees from Fig. 4.1 that

$$E = \frac{M_t}{M_t + M_b} E_{\text{lab}}, \qquad (4.1)$$

in which M_t is the mass of the target nucleus at rest in the laboratory frame of reference, and M_b is the mass of the bombarding particle. In the relativistic region one has for the sum of the kinetic energies of the two particles in the center of mass coordinate system:

$$E = [(M_t + M_b)^2 c^4 + 2 M_t c^2 E_{\text{lab}}]^{1/2} - (M_t + M_b) c^2. \qquad (4.1a)$$

4.2. Resonance Processes

For many reactions the excitation function for the cross-section $\sigma(E)$ exhibits a number of more or less pronounced maxima. This behavior is particularly evident for low-energy neutron scattering on intermediate A nuclei, and for neutron capture by heavy nuclei ((n, γ) reaction). Often the cross-section changes by as much as a factor of a thousand in passing through maxima and minima (cf. Figs. 9.1a, b). The characteristic maxima in the excitation functions of these reactions can be understood by picturing the reaction as a *resonance process*. Indeed, the general theory of nuclear reactions, at least for energies of the incident particle which are a couple of Mev or less, may be usefully organized about the concept of resonance. (Chapter 9.)

In a general reaction $X(a, b)Y$, the nuclear system consisting of X and a (or equivalently of Y and b) is called the compound system of the reaction. Thus F^{20} is the compound nucleus for the $F^{19}(n, p)O^{19}$, $F^{19}(n, \alpha)N^{16}$, $O^{18}(d, \alpha)N^{16}$, etc. reactions. The compound nucleus will have, in general, a ground state in which the particle a is bound, low-excited states which may decay only by gamma-radiation, and excited states the energy of which is greater than the separation energy of a. These latter states may decay by gamma-radiation, emission of a, and possibly by the emission of other particles. They are, of course, not stationary, but may have lifetimes long compared to the time of traversal of the particle a across the nucleus. They are called quasi-stationary states or "resonance levels." In such states the energy is not entirely sharply defined; the energy uncertainty Γ is related to the lifetime of the state by $\Gamma\tau = \hbar$.

If in the bombardment of X by a the energy of the system is close to the energy (average energy) of one of the quasi-stationary states of the compound system, the collision may be described as a two-step resonance process. The incident particle is first captured by X to form the quasi-stationary state; this state subsequently decays either by the re-emission of the particle a, or by γ-radiation, or if there is sufficient energy, by the emission of other particles. The probability of the capture of the incident particle varies in a characteristic manner with the difference between the energy E of the colliding systems, and the energy E_λ of the resonance level. The total cross-section for the process is proportional to the capture probability. This probability varies with energy roughly as $[(E - E_\lambda)^2 + (\Gamma/4)^2]^{-1}$, where Γ is the energy width of the resonance level. As the energy of the incident particle is varied over a range covering the resonance energy, the cross-section goes through a characteristic resonance maximum.

The second part of the resonance reaction process, the disintegration of the intermediate state, is governed by probability laws. The emission of each of the possible products of the collision has a certain probability characteristic of the product and of the resonance level; the different emission processes compete with each other. The transition rate for each decay process may be expressed as an energy width; this width divided by $\hbar$ is the probability of the decay per second. The energy widths associated with the different modes of decay of the resonance level are called the partial widths of the level. Thus if the partial widths for proton, neutron, and γ ray emission from a particular level are Γ_p, Γ_n, Γ_r (and no other processes are energetically possible), the probability that the disintegration eventually yields a neutron is Γ_n/Γ where $\Gamma = \Gamma_p + \Gamma_n + \Gamma_r$ is the total width of the compound state.

The essential quantities governing the behavior of cross-sections when resonance phenomena are of importance are the level positions and level widths. *On the average*, the separation between adjacent resonance levels decreases with increasing energy of excitation. Similarly, the spacing of adjacent levels decreases with increasing A. The level widths increase on the average with the excitation energy; they decrease with increasing magnitude of the electrostatic barrier, and depend also on a number of other energy-dependent factors (Chapter 9). Separations between neighboring levels, and level widths, fluctuate markedly from level to level, but separations and widths averaged over several levels appear to be relatively smoothly-varying functions of the excitation energy and of A.

The effect of the electrostatic barrier manifests itself in a reduction of the Γ for the process inhibited by the barrier. Thus the probability of an $X(n, p)Y$ process is decreased by the electrostatic barrier which inhibits the emission of the proton by decreasing Γ_p, and hence the probability Γ_p/Γ that the second process be the emission of the proton. The probability of the formation of the quasi-stationary state of the compound nucleus is not materially affected by the barrier in this case. In the $X(p, n)Y$ process, on the other hand, the approach of the proton to the target nucleus is inhibited by the barrier; in this case the cross-section for the formation of the quasi-stationary state is diminished. The barrier-penetration factor in the partial width decreases in importance as the energy of the bombarding particle increases. If the energy of the proton is sufficient to cross the barrier freely, the proton widths will be generally similar to neutron widths for the same energy.

The basic concept of the compound nucleus model is the intermediate state of relatively long lifetime. If the lifetime of any intermediate state that can be formed is so short that the corresponding width $\Gamma = \hbar/\tau$ exceeds the level spacing, the resonances begin to coalesce and the compound nucleus model ceases to be useful. This occurs for heavy nuclei at about one Mev above the threshold energy for neutron emission, and for lighter nuclei at a few Mev above that limit (cf. the Table of Nuclear Reactions at the end of this chapter). The width Γ of the levels is usually determined by the energy with which a *neutron* can be emitted, because the Coulomb barrier reduces the probability for the emission of charged particles, even if they could be emitted with a higher kinetic energy.

4.3. Direct Processes

The mechanism of nuclear reactions becomes rather simpler in the energy region above individual distinct resonances because in the absence of such resonances the behavior of the nucleus becomes more nearly classical. Two energy regions may be distinguished in such a case. If the velocity of the incident particle is so low that the nucleons of the target nucleus rearrange very often while the incident nucleon traverses the nucleus, it will be possible to represent the effect of the target nucleus by an average potential. The corresponding model is called the "optical model" because of the analogy with the behavior of a refracting medium. At even higher energy of the incident particle the interaction of the nucleons of the nucleus can be neglected, and the nucleus treated as an aggregate of free particles. Both models are called direct interaction models to express the absence of the compound state as an intermediary between initial and final states. In some cases the direct interaction may become significant between the resonances also at low energies.

The direct interaction models are most useful to describe reactions induced by deuterons. The deuteron as a charged particle is repelled by the target, and this makes the formation of the compound nucleus rather improbable. However, the repulsion does not prevent the direct process equally effectively because the deuteron is a rather loose structure with a low binding energy in which relatively large neutron-proton separations are not improbable. As a result the neutron may enter the nucleus and be absorbed by it, while the proton remains relatively far from it, not exposed to the electrostatic repulsion to the same extent as if both particles had to enter the nucleus. The resulting process is called "stripping" because the neutron is stripped off the proton by the collision. Similar processes play a role also in (d, n) reactions and in the inverse "pick-up" reactions (n, d), (p, d).

The stripping and pick-up processes are only examples of the more general class of direct processes. Common to the theory of all these processes is the fact that the effective interaction between target and incident particle takes place close to the surface of the nucleus, and the fact that it can be treated as a perturbation—a treatment which would yield incorrect results in the resonance region. The theory of the direct processes will be considered in Chapter 10.

For heavy nuclei $(A > 90)$ charged particle reactions are inhibited by the large electrostatic barrier if the incident energy is below about 5 Mev (the potential energy at the top of the barrier is of the order $0.4\ A^{2/3}$ Mev, i.e. about 8 Mev for $A = 100$, 14 Mev for $A = 200$). Relatively low-energy charged particles may nevertheless be employed for excitation of heavy nuclei by the process of electric or Coulomb excitation. This process does not depend on specific nuclear interactions. Passage of the incident particle by the nucleus—the barrier prevents the incident particle from reaching the surface of the target nucleus—subjects the nuclear system to a rapidly changing electromagnetic field. This field gives rise to the transition. Thus the electric excitation is a particular kind of photoeffect. The function of the incident particle is to provide intense electromagnetic radiations in the neighborhood of the nucleus.

4.4. Table of Most Important Reactions

In the accompanying table the characteristic nuclear reactions are presented. The table is separated into sections depending on the mass number of the target, the nature of the incident particle, and the magnitude of the incident kinetic energy. These divisions reflect the overriding importance of energy availability in determining the properties of reactions. The nature of a reaction is specified in the table by indicating the reaction product. The reactions are listed in the order of their cross-sections. Reactions with cross-sections less than one percent of the leading cross-section are not shown. In scattering processes elastic and inelastic scattering are distinguished by the abbreviations el and inel. The observation of individual resonance levels is indicated by res. (Elastic scattering occurs, of course, in all cases and is only indicated in connection with resonance properties).

The importance of barrier effects is shown by absence of reactions involving charged incident particles in the low-energy range for the intermediate mass nuclei, and in the low- and intermediate-energy ranges for the heavier nuclei.

As mentioned before, for high energies more than one reaction product becomes possible. Frequently these reactions may be considered as a succession of two-body events. Thus the reaction $X(\alpha, np)Y$ may be written $\alpha + X \rightarrow Y' + n$, $Y' \rightarrow Y + p$.

No corresponding chart can be given for the nuclei with $A < 25$. In this region the binding energy is an irregular function of A; since the character of a reaction depends so much on the availability of energy, the collisions with light target nuclei must be considered as special cases.

Nuclear reactions can be used to produce radioactive elements. They often lead also to excited states of nuclei of longer or shorter lifetimes, the study of which is of great importance for the understanding of nuclear structure.

TABLE OF NUCLEAR REACTIONS

Incident Particle	Intermediate Nuclei ($30 < A < 90$)				Heavy Nuclei ($A > 90$)			
Energy of Incident Particle	n	p	α	d	n	p	α	d
Low 0–1 kev	n(el) γ (res)	no appreciable reaction	no appreciable reaction	no appreciable reaction	γ n(el) (res)	no appreciable reaction	no appreciable reaction	no appreciable reaction
Intermediate 1–500 kev	n(el) γ (res)	n γ α (res)	n γ p (res)	p n	n(el) γ (res)	Very small reaction cross-section	Very small reaction cross-section	Small reaction cross-section
High 0.5–10 Mev	n(el) n(inel) p α (res for lower energies)	n p(inel) α (res for lower energies)	n p α(inel) (res for lower energies)	p n pn $2n$	n(el) n(inel) p γ	n p(inel) γ	n p γ	p n pn $2n$
Very high 10–50 Mev	$2n$ n(inel) n(el) p np $2p$ α three or more particles	$2n$ n p(inel) np $2p$ α three or more particles	$2n$ n p np $2p$ α(inel) three or more particles	p $2n$ pn $3n$ d(inel) tritons three or more particles	$2n$ n(inel) n(el) p pn $2p$ α three or more particles	$2n$ n p(inel) np $2p$ α three or more particles	$2n$ n p np $2p$ α(inel) three or more particles	p $2n$ np $3n$ d(inel) tritons three or more particles

CHAPTER 5

Two-Body Systems; Interactions Between Nucleons

5.1. Internucleon Forces

Our most detailed information on the nature of the interactions between nucleons is derived from a study of the behavior of two-body systems, in particular from study of properties of the deuteron and from the analysis of neutron-proton and proton-proton scattering experiments. The two-body systems play a special role since only for them can the theoretical consequences of an assumed interaction between the nucleons be calculated with precision. The problem is to invent neutron-proton and proton-proton interactions in terms of which the observed properties of these two particle systems can be derived theoretically. (We have no direct information on the neutron-neutron interaction since two neutrons form no bound system and, as yet, it is impossible to perform neutron-neutron scattering experiments).

The data which must be correlated by an assumed n-p interaction are very extensive. The neutron-proton system has a bound state, the normal state of the deuteron H^2, with a binding energy 2.23 mev, spin 1, magnetic moment 0.8573 nuclear magnetons, and quadrupole moment of 2.74×10^{-27} cm^2. Neutron scattering on protons has been studied extensively over a broad range of energies; it has been investigated in considerable detail from thermal energies to about 18 mev and, in somewhat less detail, up to several hundred mev. In addition, the cross-sections for photo-dissociation of the deuteron (at various γ-energies), for the capture of neutrons by protons, and for the scattering of neutrons by ortho- and para-hydrogen at very low energies ($\sim$.01 ev) are known.

It is possible to formulate a neutron-proton interaction from which these observed properties of the neutron-proton system can be calculated. The main features of the interaction which provides a fit with the observed data are:

1. The interaction is of short range—1 to 2×10^{-13} cm. This is required by the observed isotropy of the neutron-proton angular distribution at energies up to $\sim$ 10 mev, by the sensitive experiments on the scattering by ortho- and para-hydrogen, as well as by other data.

2. The interaction is spin dependent, i.e. the force depends on the relative orientation of the spins of the two particles. This conclusion was forced by the behavior of the scattering cross-section at low neutron

energies (~1 ev), and is most clearly indicated by the large difference in the cross-sections of ortho- and para-hydrogen for thermal neutrons. These last cross-sections also give a rather accurate value for the strength (effective depth of the potential) of the neutron-proton interaction for the singlet state (anti-parallel spins). The binding energy of the deuteron provides a measure of this strength only for the triplet state (parallel spins). The cross-sections for ortho- and para-hydrogen also show that the neutron spin is $\frac{1}{2}$.

3. The neutron-proton interaction depends not only on the orientation of the proton and neutron spins with respect to each other, but also on their orientation with respect to the line passing through the two nucleons. An interaction of this character is required to explain the observed quadrupole moment; this moment indicates that the proton distribution in the ground state of the deuteron is not spherically symmetrical, but is elongated in the direction of the total spin of the system. The simplest interaction which depends on the orientation of the spins σ_1, σ_2 of neutron and proton relative to the radius vector $\mathbf{r}$, which specifies the separation of these particles, is

$$V_T = J_T(r)\left\{\frac{3(\sigma_1\cdot\mathbf{r})(\sigma_2\cdot\mathbf{r})}{r^2} - (\sigma_1\cdot\sigma_2)\right\} = J_T(r)S_{12}. \qquad (5.1)$$

$J_T(r)$ specifies the radial dependence of the interaction. V_T is called the tensor interaction.

4. There is some evidence that the nuclear interaction is repulsive at very small separations of the nucleons. A particular potential which was originally proposed on the basis of the meson theory of nuclear forces and which gives a reasonably accurate description of two body systems is Levy's interaction. This is strongly repulsive at very short distances ($r < 0.55 \times 10^{-13}$ cm). Outside of this range it consists of three parts. The first part is a so-called exchange interaction:

$$V_1 = -0.55\frac{\hbar c}{r}e^{-r/a}(\mathbf{P} + (1/3)\sigma_1\cdot\sigma_2(1 - \mathbf{P})), \qquad (5.2)$$

where $a = 1.40 \times 10^{-13}$ cm is the Compton wave-length of the π meson. $\mathbf{P}$ is the so-called Majorana exchange operator: it gives a factor 1 if the wave function is symmetric with respect to the interchange of the space coordinates of the interacting particles; it gives a factor -1 if the wave function is antisymmetric with respect to such an interchange. In the case of a two-body system $\mathbf{P}$ is 1 if the orbital angular momentum is even (S, D, G, etc. states), $\mathbf{P} = -1$ if the orbital angular momentum is odd (P, F, etc. states).

The second part of Levy's interaction is a tensor force:

$$V_2 = -0.55\frac{\hbar c}{r}e^{-r/a}(1 + 3a/r + 3a^2/r^2)S_{12}(2\mathbf{P} + 1), \qquad (5.3)$$

where S_{12} is given by the curly bracket of (5.1). The last part of the interaction is given by an ordinary potential; it does not have exchange character nor does it depend on the orientation of the spin:

$$V_3 = -\frac{\hbar c a}{r^2}\left(K_1\left(\frac{2r}{a}\right) + .05\left[K_1\left(\frac{r}{a}\right)\right]^2\right). \tag{5.4}$$

The K_1 is the Hankel function as defined, e.g. in Watson's *Theory of Bessel Functions*. The derivation from meson theory of this last part of the nuclear interaction has been questioned. From the point of view of the interpretation of the two-body systems, an attractive interaction similar to (5.4) is undoubtedly needed. The whole two-body interaction is, therefore, according to Levy:

$$V = V_1 + V_2 + V_3 \quad \text{for} \quad r > 0.38a \tag{5.5}$$

V large and positive for $r < 0.38a$.

Levy's potential which is given above will not be used below for actual calculations. It is included here only to illustrate the attempts to obtain an expression for the nuclear interaction. Two of the constants in Levy's potential (the range $0.38a$ of the hard core and an overall multiplicative constant) were so determined as to give the observed values for the binding energy of the deuteron and for the neutron-proton scattering at low energies. In addition to these, the potential reproduces the observed behavior of the neutron-proton and proton-proton scattering up to about 30 mev and gives a fairly accurate value for the quadrupole moment of the deuteron. Actually, the majority of nuclear phenomena depend only on the interaction of particles with very low angular momenta about their common center of mass. In particular, unless the energy is well above 18 mev, the two-body system's interaction is restricted to angular momenta 0 and 1. Hence, only the low angular momentum parts of the potentials can be tested by experiments below about 18 mev on two-body systems.

Another potential which has also been quite successful was obtained from an analysis of the two-particle experiments with which the present section is concerned. It is:

$$-\frac{0.25\hbar c}{r}(1 - 0.04(\mathbf{d}_1 \cdot \mathbf{d}_2))e^{-1.2r/a}$$
$$- \left[\frac{0.26\hbar c}{r} S_{12} e^{-0.92r/a}(0.38 + 0.62\mathrm{P})\right]. \tag{5.6}$$

The fact that (5.6) differs from the Levy potential quite substantially, and that both give a fair account of the two-particle system (though not of the properties of heavier nuclei) shows that the nuclear interaction cannot be unambiguously determined from the known properties of two-particle systems.

The proton-proton system has no bound state. The angular distributions obtained from proton-proton scattering experiments have been measured at a large variety of energies between 300 kev and 30 mev and at some higher energies up to 340 mev. The analysis of the data for energies below 5 mev is relatively simple since the scattering depends only on the proton-proton (p-p) interaction in the singlet S state. The triplet S state is absent because the wave function must be antisymmetric. The data may be fitted by the assumption that the p-p interaction can be obtained from the n-p interaction in the singlet S state by the addition of the electrostatic term e^2/r. Since the properties of mirror nuclei suggest that the n-n and p-p interactions are equal (again aside from electrostatic effects), equivalence of the p-p and n-p interactions (in the ^1S state) has led to the assumption of *charge independence* of nuclear forces, i.e. that all interactions between nucleons are equal, apart from electrostatic effects. This assumption has been incorporated also into meson theory and plays a fundamental role there.

The high-energy scattering cross-sections for the p-p and the n-p systems differ considerably. Such a difference is to be expected because the n-p system contains states which the p-p system, on account of the exclusion principle, cannot contain. However, attempts to explain the difference between p-p and n-p cross-sections on this basis have not yet been fully successful.

5.2. Saturation Properties and Internucleon Forces

If one assumes that the potential energy of a nuclear system is the sum of two-particle interaction energies, then the fact that the binding energy per-particle of nuclei is roughly constant—the saturation property—leads to important theoretical restrictions on the nature of the interactions between nucleons. The interaction cannot be attractive throughout and of the ordinary type, as this would not lead to saturation. If the interaction were of this nature, all particles would tend to be in the range of each other. Hence, the nuclear volume would be constant rather than proportional to A, and the binding energy would be proportional to A^2 rather than to A, as observed. The saturation property can be obtained by a mixture of ordinary and Majorana forces of the form $V = J(r)\,[\alpha + \beta \mathbf{P}]$ with $J(r)$ negative for all r only if $\beta > 4\alpha$, i.e. if the Majorana force dominates. However, interactions which satisfy the saturation requirements predict high-energy n-p and p-p scattering distributions in disagreement with the scattering data. It appears, consequently, that it is necessary to assume either that $J(r)$ is not negative throughout (cf. the Levy potential which is positive for $r < 0.55 \times 10^{-13}$ cm), or that the interaction between nucleons within nuclear matter is different from the interaction between isolated pairs of nucleons, i.e. that many-body forces (as distinguished from the usual two-body interactions) are of importance. It is believed at present that the former effect, the repulsive core, is more important.

5.3. Charge Independence of Nuclear Forces: The Isotopic or Isobaric Spin Quantum Number

The hypothesis of the charge independence of nuclear forces has proved increasingly fruitful. If nuclear interactions are in fact charge independent, i.e. the forces between nucleons are the same for neutrons and protons, then the states of a nuclear system may be characterized by an additional quantum number T—the so-called isotopic or isobaric spin quantum number.

The significance of T is analogous to that of the spin quantum number S for atomic states. The states of the atom are, according to Pauli's principle, wholly antisymmetric with respect to the exchange of space *and* spin coordinates of the electrons. As a consequence the *symmetry character* of an atomic state, with respect to exchange of space coordinates only, is completely defined by the quantum number S, the total spin angular momentum number of the state: S determines the symmetry character with respect to the exchange of the spin variables and the symmetry character with respect to the exchange of space coordinates must be opposite thereto. If the atomic Hamiltonian is independent of the spin variables (and this is approximately true for light atoms), then S is a constant of motion, and different states with different values of S are characterized by different *spatial* symmetries.

A similar argument can be applied to nuclear states. A new variable—the isotopic spin variable—may be introduced for nucleons. This variable τ plays a role similar to that of the electron spin variable. It distinguishes between neutrons and protons just as the electron spin variable is used to distinguish electrons with spins along or opposed to the z axis. With this additional variable the states of a nuclear system must be antisymmetric with respect to the simultaneous exchange of the space, ordinary spin, and isotopic spin coordinates of any two particles. The quantum number T bears the same relation to the isotopic spin variable τ that S does to the electron spin variable s. The symmetry character of a state with respect to exchanges of *space and ordinary spin* variables is uniquely defined if the state is characterized by a definite value of the total isotopic spin quantum number T. If the nuclear Hamiltonian is independent of the isotopic spin variable, i.e. if the forces are the same for protons and neutrons, then T is a constant of motion, and has well defined values for the stationary states of nuclei.

The isotopic spin quantum number T will regulate transitions between nuclear states in much the same way as do the angular momentum quantum numbers. Conversely, recent work on nuclear reactions furnishing evidence that T is a constant of motion, at least for the lighter nuclei, supports the assumption of the charge independence of nuclear forces.

In the atomic case there are $2S + 1$ independent states belonging to the same quantum number S with different values of the z component S_z of the spin angular momentum. S_z determines the difference between the number of electrons with spin up and the number with spin down. If the atomic Hamiltonian is spin independent, all these states have the same energy. Similarly, there are $2T + 1$ independent nuclear states associated with isotopic spin quantum number T belonging to different values of T_ζ. ($T_\zeta = T, T - 1, \cdots, -T$). This latter variable measures the difference between the number of neutrons and protons in the nucleus and has the value $(N - Z)/2$ for a nucleus with N neutrons and Z protons. The different T_ζ associated with a particular T describe states of several isobaric nuclei; the dependence of all these states on the space and ordinary spin variables of the nucleons is the same. If the nuclear Hamiltonian is charge independent, these states have the same energy. Hence, while the quantum number S stipulates the equality of the energy of those states of a given atom which differ in the component of the spin angular momentum in a given direction, the quantum number T postulates equality of the energy of states of several isobars: those united into a T multiplet. This is illustrated in Fig. 5.1a. Two states of a T multiplet differ only by neutrons being substituted in one of these states for some of the protons in the other state, or vice versa. Thus, naturally, the total angular momentum J and the parity is the same for all states of a T multiplet.

Fig. 5.1a does not give an accurate picture of the energy values of isobars; it must be corrected for the difference between the interaction of two protons and that of two neutrons or of a proton-neutron pair. According to the hypothesis of charge independence, the latter two interactions are equal while the proton-proton interaction differs from them by the electrostatic repulsion of the protons. As a result of this interaction the energy is not independent of T_ζ; the members of the T multiplet with higher T_ζ (fewer protons) have lower energy. This is illustrated in Fig. 5.1b. Since the electrostatic potential within a nucleus can hardly be expected to depend critically on the nuclear state, the dependence on T_ζ is very nearly the same for all T multiplets. Hence the slope of the lines connecting the members of a T multiplet is very nearly the same throughout Fig. 5.1b. It follows from this picture that the T value of the normal state of a nucleus which is stable against positron emission and K-capture is equal to its T_ζ value, i.e. equal to $\frac{1}{2}(N - Z)$ for that nucleus. Clearly, T cannot be smaller than the T_ζ

of the stable isobar. If T were larger, the T multiplet would have a state with a larger T_ζ. The total energy of this state would then be smaller and the original nucleus would be able to decay into it by positron emission or K-capture. It follows, for instance, that $T = 0$ for the normal states of H_1^2, He_2^4, Li_3^6, B_5^{10}, C_6^{12}, N_7^{14}, O_8^{16}; and $T = \frac{1}{2}$

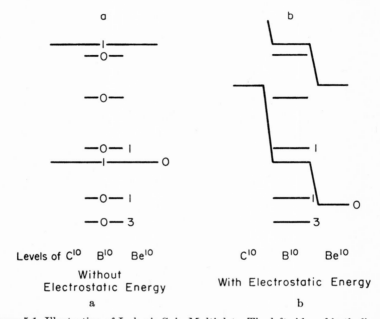

Figure 5.1. Illustration of Isobaric Spin Multiplets. The left sides of both diagrams give the levels of a nucleus with $T_\zeta = -1$ ($N - Z = -2$), the centers give the levels of the isobar with an equal number of neutrons and protons ($T_\zeta = 0$); the right sides, the levels of the isobar with $T_\zeta = 1$ ($N - Z = 2$). The energies of the levels are plotted on a vertical scale. The J values (spins) of the levels are shown to the right. The numbers in the middle (Fig. 5.1a) indicate the T values of the multiplets. The diagram on the left would be valid if the electrostatic energy could be disregarded. In this case the binding energies of the levels of the $T_\zeta = \pm 1$ isobars would be equal, and the $T_\zeta = 0$ isobar would have levels at the same binding energies. The three states with the same binding energies form an isobaric spin triplet. The nucleus with $T_\zeta = 0$ has further levels which are isobaric spin singlet levels ($T = 0$). The diagram on the right illustrates the change in the binding energy, introduced by the electrostatic forces. This diagram actually gives the low-lying levels of C^{10} (left side), of B^{10} (middle), and of Be^{10} (right side of the diagram).

for the normal states of all odd A nuclei up to Cl_{17}^{35}. The T value of the normal state of Cl_{17}^{37} is, on the other hand, $\frac{3}{2}$; that of O_8^{18}, Ne_{10}^{22}, Mg_{12}^{26}, etc. is 1.

The normal states of mirror nuclei such as Li^7 and Be^7 are the two members of the same isotopic spin doublet; their binding energies differ only as far as electrostatic effects are concerned. We have used this fact in Chapter 3 to calculate the difference between the electrostatic energies of mirror nuclei, and hence to estimate their size. The normal state of Be_4^{10} ($T = 1$) is stable against K-capture or positron emission (it decays by electron emission). Hence $T = 1$ for this state. The other members of this T multiplet are the normal state of C_6^{10} ($T_\zeta = -1$) and the 1.74 mev excited state of B_5^{10} ($T_\zeta = 0$). The binding energies of these states differ from that of the normal state of Be_4^{10} only by the electrostatic energy and can easily be calculated therefrom. Many similar examples are known.

In addition to the consequences which one can derive from the concept of the T multiplet for the binding energies of the states of a T multiplet, a number of selection rules follow from the validity of the quantum number T. These will be stated here without derivation.

Collision of two particles with isotopic spin T_1 and T_2 can give only compound states with T between $|T_1 - T_2|$ and $T_1 + T_2$. Conversely, disintegration of a compound state with a given T value always gives particles in such states that their isotopic spins T'_1 and T'_2 can form a triangle with T, i.e. that $|T'_1 - T'_2| \leq T \leq T'_1 + T'_2$. Even if the reaction does not proceed via a compound state, the reaction is "forbidden" if the intervals $(|T_1 - T_2|, T_1 + T_2)$, and $(|T'_1 - T'_2|, T'_1 + T'_2)$ have no common element. Thus, the reaction $B_5^{10} + H_1^2$ should not lead to He^4 plus Be^8 in a $T = 1$ state. Electromagnetic radiation can change T only by 0 or ± 1, and $0 \to 0$ is forbidden for dipole transitions (naturally, electromagnetic radiation does not change T_ζ). Similarly, only $\Delta T = \pm 1$ is possible in allowed β-transitions ($\Delta T_\zeta = \pm 1$ in this case). Experimentally, all these rules seem to be valid in the great majority of cases and there is only a handful of cases for which an explanation is still lacking.

CHAPTER 6

Nuclear Models A. The Uniform Model

6.1. General Remarks

Theoretical investigation of the properties of nuclear systems for $A > 2$ is made difficult by two features. First, we do not as yet have a clear conception of the nature of nuclear forces, although it becomes increasingly clear that they are rather complicated. Second, even if we assume forces of simple form but of short range, there appears to be no simple approximation procedure by means of which reasonably accurate solutions of the Schrödinger equation for many nucleon systems can be obtained. The situation for nuclei is in sharp contrast to that for atoms. For the latter systems the major interactions are well understood and substantiated; the nature of the forces is such that it is possible to justify, for the calculation of atomic wave functions, a rather simple but yet very powerful approximation procedure—the so-called shell model of the atom. As yet no nuclear model exists which has anything like the range of validity of the shell model for the atom.

Recently, it has become possible to calculate at least the properties of "nuclear matter" from an assumed interaction between pairs of nucleons. Such properties are, for instance, the energy density in a nucleus far from the nuclear surface, or the probability for a given distance between two nucleons under the same conditions. These properties are analogous to volume properties of a solid or of a liquid, as contrasted with its surface properties. The significance of this progress, theoretical and practical, is hardly foreseeable. It should be recalled, nevertheless, that even in a nucleus as heavy as Pb, more than half of the nucleons are at the surface, and also that the nucleons in the outer layers of the nucleus have more influence on its properties than the nucleons inside.

The concept of a "model" as we are using it here must be understood in a very broad sense. Any set of simplifying assumptions, physical or mathematical, by means of which the characteristics of nuclear systems may be computed to some degree of approximation may be called a model. So far it has not been possible to justify by strict reasoning the simplifying assumptions of any of the proposed models. A model may be evaluated for the most part only by an investigation of the extent of its success and failure in describing the observed properties of nuclei. Even when a model is partially successful, it is often difficult to estimate to what extent its success reflects the accuracy of the model. The more or less extensively studied models are briefly reviewed in this and the following two chapters.

6.2. Powder and Shell Models

Two broad classes of models have been proposed for nuclei. The powder models assume that the nuclear wave functions are very complicated and resemble the elementary chaos encountered in statistical mechanics. The shell models, on the contrary, liken the nuclear structure to a planetary system or to the electrons in atoms, which move, at least in first approximation, independently of each other and are arranged in regular shells. Clearly, the two pictures are rather contradictory and their ranges of validity must be different. The shell or independent particle models are most nearly valid for the normal and low-excited states of nuclei with closed shells, or with only one particle missing from closed shells or one particle present outside of closed shells. These are the conditions under which the independent particle model for the electronic shells is most accurate, and they constitute the best conditions for this model also in nuclei. The powder model, on the other hand, can be expected to be valid for more highly-excited states, or if the nuclear constitution is far removed from that of a closed shell. In both cases there are many close-lying states of the nucleus and the interaction between these states creates conditions which resemble chaos.

The question naturally arises whether the powder model permits the calculation of any nuclear properties and, indeed, the powder model suffers from the paucity of conclusions which it permits. It has been, in its unadulterated form, useful principally in the theory of nuclear reactions which deals, naturally, with nuclear systems at rather high excitations. Even for these, it seems to yield to less extreme models. As far as stationary states are concerned, the conclusions from the powder model are much more limited than those of statistical mechanics dealing with molecular chaos.

Naturally, the general conservation laws of quantum mechanics are valid for all stationary states, no matter what the model from which they arise: each stationary state will have a total angular momentum J and a parity. The isobaric spin quantum number T will be a valid concept for not too highly-excited states. However, more detailed results concerning the properties of stationary states can be derived only by assuming that the nuclear wave function is largely determined by the spin independent forces alone. The consequences of this assumption are investigated in the so-called supermultiplet theory. The conclusions are applicable not only to the powder model but to all nuclear models, such as the L-S coupling shell model, which neglect the spin dependent forces at least in first approximation.

The validity of the assumption that the spin dependent forces do not substantially influence the wave function is more severely limited than the validity of the isobaric spin concept. The electrostatic forces are both weaker and less effective in influencing the wave function than the spin dependent forces. It seems that the validity of the supermultiplet theory's conclusions is confined, in any case, to light elements. Their validity for these will be discussed in more detail in the next section.

Nuclear theory was late in recognizing the value and usefulness of the independent particle model for nuclei. The reason for this is that the short range of the nuclear forces is hardly compatible with the independence of the motion of nucleons and their ability to complete several revolutions in individual orbits undisturbed. It will be pointed out again in section 7.8 that the independent particle model is valid for nuclei in a much more subtle sense than for atoms. However, the consequences of the model are hardly affected by the refinements. They encompass a remarkably wide field of applications.

6.3. Supermultiplet Theory

This is concerned with the consequences of the assumption that one can obtain a reasonable wave function for the nucleus by assuming only spin and isotopic spin independent interactions between the nucleons. The electrostatic and the tensor forces are assumed to have no influence, or only very little influence on the wave function; and thus the state is presumed to have nearly the same form as if only ordinary and Majorana exchange forces were present. However, when calculating the total energy of the nucleus, the energy of the spin dependent and electrostatic forces is added to the energy contribution of the ordinary and Majorana forces. Expressed in more technical language, the spin dependent and electrostatic forces are taken into account only in first approximation. Some attempts have been made to calculate the effect of the tensor forces also in second approximation. To illustrate this by an example: the uniform model assumes that the density of the protons is constant throughout the nucleus. Hence, the total electrostatic energy is (apart from exchange effects) that of a uniformly charged sphere: $0.6Z(Z - 1)e^2/R$, where R is the nuclear radius. Actually, the electrostatic forces themselves affect the density of the protons in the nucleus so that this is not uniform. As a result, the electrostatic energy differs from the above value but this difference is neglected as a second-order effect.

Both the total spin S and the isotopic spin T are good quantum numbers in supermultiplet theory. The validity of the latter quantum number was discussed in the preceding chapter and found to be adequate; the validity of S as a good quantum number—which corresponds to the Russell-Saunders coupling in atomic theory—is more open to question. The validity of both quantum numbers S and T implies that, in first approximation, $(2S + 1)(2T + 1)$ states of a set of isobars have the same energy: the S_z and T_ζ values of these states range independently from $-S$ to S and from $-T$ to T, respectively.

Rotations in ordinary and isotopic spin space are not the only symmetry operations: one can further interchange, *for instance*, the spin coordinates with the isotopic spin coordinates without affecting the Hamiltonian of the system. As a result of the last symmetry operation, states with isotopic spin S and ordinary spin T will coincide, in the approximation considered, with states with isotopic spin T and ordinary spin S. The set of states the energies of which coincide in first approximation is much larger than in ordinary spectroscopy. It is called, therefore, a supermultiplet; it needs for its characterization, three numbers: P, P', P''. The mathematical theory of the supermultiplets is much more involved than that of the multiplets of ordinary spectroscopy (or the mathematically identical theory of T multiplets of the preceding section). The meaning of the numbers P, P', and P'' is as follows: P is the largest T_ζ of any state contained in the supermultiplet, P' is the largest S_z of any state with $T_\zeta = P$. The last quantum number P'' plays a subordinate role; it is the largest value of

$$\tfrac{1}{2}(\sigma_1\tau_1 + \sigma_2\tau_2 + \cdots + \sigma_A\tau_A) \tag{6.1}$$

for any state with $T_\zeta = P$, $S_z = P'$. It follows that P, P' and P'' are all integers for even A, half integers for odd A.

The largest T multiplet contained in the supermultiplet is clearly the largest T_ζ, i.e. is equal to P. The largest S value of any $T = P$ state is the largest S_z of any $T_\zeta = P$ state, i.e. it is equal to P'. Hence, the largest T, S combination is $T = P$, $S = P'$. Since the supermultiplet contains, along with a T, S combination, also an S, T combination, and since P is the largest T contained in the supermultiplet, $P \geq P'$, and one can show further that

$$P \geq P' \geq |P''|. \tag{6.2}$$

While P and P' are always positive, P'' can be also negative. It may also be proved that

$$P + P' + P'' + \tfrac{1}{2}A = 2n \tag{6.2a}$$

is always an even number.

It is important for the understanding of ordinary spectroscopy that among all the spin functions of N spin variables, $S = \tfrac{1}{2}N, \tfrac{1}{2}N - 1, \tfrac{1}{2}N - 2, \cdots$; the last one, i.e. either $S = 0$ or $S = \tfrac{1}{2}$, is the most nearly antisymmetric. (This is a concept which we shall not define rigorously.) As a result, and as a consequence of the antisymmetry of the total wave function, the coordinate function associated with $S = 0$ (in the case of even N) or $S = \frac{1}{2}$ (in the case of odd N) is most nearly symmetric. The situation is quite similar for supermultiplets. The most nearly antisymmetric functions of A variable pairs $\sigma_1, \tau_1; \sigma_2, \tau_2; \cdots ; \sigma_A, \tau_A$ are those which have the smallest values of P, P', and P''. Antisymmetry is meant in this case with respect to the interchange of pairs of variables, for instance the interchange of σ_2 with σ_3 *and* τ_2 with τ_3. Again, as a result of this and of the antisymmetry of the total wave function with respect to the interchanges of all coordinates (position, spin, and isobaric spin), the part of the wave function which depends only on the position coordinates is most nearly symmetric for the smallest possible values of P, P', and P''. As an example: for $A = 2$ the functions of $\sigma_1, \tau_1; \sigma_2, \tau_2$ are antisymmetric for $(P, P', P'') = (1, 0, 0)$ but symmetric for $(P, P', P'') = (1, 1, 1)$. Similarly for $A = 3$ and 4, the spin-isobaric spin functions are completely antisymmetric in the $(\frac{1}{2}, \frac{1}{2}, \frac{1}{2})$ and $(0, 0, 0)$ supermultiplets, respectively. (For higher A there is no completely antisymmetric spin-isobaric spin function). It follows that the position coordinate function is most nearly symmetric for the lowest possible values of P, P', P''—it is completely symmetric in this case for $A = 2, 3, 4$.

The more nearly symmetric the position coordinate function is, the lower is the kinetic energy and the more negative is the potential energy due to the spin- and isobaric spin-independent ordinary and Majorana exchange forces. This is particularly clear for the Majorana interaction which has, as explained after (5.2), opposite signs for symmetric and antisymmetric position coordinate wave functions. Since the Majorana interaction is attractive for a symmetric wave function, it is repulsive for an antisymmetric one. The effect is similar, though less pronounced, for ordinary forces. These are (except for the repulsive core) attractive for both symmetric and antisymmetric wave functions. However, in the latter case the wave function must vanish when the position coordinates coincide, and remains relatively small within the range of interaction. As a result, the attractive short range ordinary potential will have a much smaller value for an antisymmetric than for a symmetric wave function. A similar argument applies also for the kinetic energy. As a result, the quantum numbers P, P', and P'' of the lowest lying supermultiplets will be as low as is consistent with the T_ζ of the nucleus in question and with the rules (6.2), (6.2a). Since the largest value of T_ζ in a supermultiplet is P, the smallest value which P can assume is T_ζ. The remaining quantum numbers P' and P'' will be as low as possible in view of (6.2), (6.2a) and never larger than 1. More explicitly, they will be:

$(P, P', P'') = (|T_\zeta|, 0, 0)$ for even-even nuclei,
$\qquad\qquad = (|T_\zeta|, \frac{1}{2}, \pm\frac{1}{2})$ for even-odd or odd-even nuclei,
$\qquad\qquad = (|T_\zeta|, 1, 0)$ for odd-odd nuclei, $(T_\zeta > 0)$,
$\qquad\qquad = (1, 0, 0)$ for odd-odd nuclei, $(T_\zeta = 0)$.

6.3 · SUPERMULTIPLETS

In the third case $P' = 1$ is required by (6.2a): $T_{\zeta} + 1 + \frac{1}{2}A = \frac{1}{2}(N - Z) + 1 + \frac{1}{2}(N + Z) = N + 1$ is even in this case, but would be odd if the 1 were replaced by a 0. A similar remark applies to the last case.

Since P' is the largest value of S_z which is compatible with the value P for T_{ζ}, one infers at once $S = 0$ for the normal state of even-even nuclei, and $S = \frac{1}{2}$ for the normal states of even-odd and odd-even nuclei. For odd-odd nuclei one will infer $S = 1$. These consequences of the supermultiplet theory cannot be compared directly with experiment because S cannot be measured by itself; only the vector sum J of the spin angular momentum S and the orbital angular momentum L is directly observable as the "spin" of the nucleus. J is always 0 for even-even nuclei so that one would have to postulate $L = 0$ for the normal states of all even-even nuclei. The postulates concerning S, while not subject to direct experimental test, do stand in direct conflict with postulates of the j-j coupling model discussed later. According to present evidence, at least for medium heavy and heavy nuclei, the wave function furnished by the j-j coupling model comes closer to the actual wave function than any wave function compatible with the supermultiplet theory. However, in our opinion the supermultiplet theory and its quantum numbers P, P', and P'' constitute a good approximation for light nuclei. This statement means that the spin dependent forces and the electrostatic interaction, though large in magnitude, do not influence the wave function itself too radically. This is confirmed also by direct calculations which take the spin dependent forces into account but which yield wave functions ninety percent of which have the supermultiplet character described above.

The great weakness of the supermultiplet theory is its lack of specificity, i.e. that it provides only some general characteristics for the wave function, but does not give a definite expression therefor. As a result, those properties of the nuclei which do not follow from general symmetry arguments cannot be calculated on the basis of the supermultiplet theory alone. The two most important results of the theory which can be compared with experiment directly concern the magnitude of the allowed β-decay matrix elements and the general trend of nuclear binding energies. With respect to β-decay, supermultiplet theory provides a distinction between favored and unfavored transitions, with the former having, for the same decay energy, much shorter lifetimes than the latter. The favored transitions should occur only between states which belong to the same supermultiplet—a conclusion which seems to be confirmed by present experimental evidence.

Concerning the general trend of nuclear binding energies, the supermultiplet theory leads to formulas for the binding energy similar to Weizsäcker's expression (2.1). In particular, the decreased binding energies of odd-odd nuclei, and of even-odd and odd-even nuclei, as compared with the binding energies of even-even nuclei (i.e. the $\delta/A^{3/4}$ terms in (2.1)), result from the P', $P'' = 1,0$ and P', $P'' = \frac{1}{2}$, $\pm\frac{1}{2}$ in the symmetry symbols for the corresponding types of nuclei. One also obtains in this way an interpretation of the "symmetry term" $U_t T^2/A$ and a connection between this term and the $\delta/A^{3/4}$ term which is, for not too heavy nuclei, well confirmed by the values of U_t and δ. The special role of the odd-odd nuclei with $T = 0$ deserves further comment. The only stable odd-odd nuclei (H^2, Li^6, B^{10}, N^{14}) belong to this type. Their supermultiplet (1, 0, 0) gives two S, T combinations: $S = 1$, $T = 0$ and $S = 0$, $T = 1$. The former gives the lowest state of the aforementioned odd-odd nuclei, which are therefore in the triplet state; the latter combination gives the normal states of the adjoining even-even nuclei (He^6, Be^6; Be^{10}, C^{10}; C^{14}, O^{14}) as well as excited states of the odd-odd nuclei. All these are singlets, in fact, having $J = 0$. Neglecting all spin dependent and electrostatic forces, their energies are therefore exactly equal. The spin dependent forces will depress some of the triplet states; the electrostatic forces, the states with larger T. For lower A the former predominate, hence H^2, Li^6, B^{10}, and N^{14} are stable. However, as A and hence Z increase, the electrostatic forces begin to prevail and the normal state of F^{18} (which is a triplet state) lies higher than the normal state of O^{18} which is a singlet but has less electrostatic energy.

The (1, 0, 0) supermultiplet is the only one which gives the normal states of two types of nuclei: those of odd-odd $T_\zeta = 0$ and of even-even $T_\zeta = \pm 1$ nuclei. This explanation of the stability of certain odd-odd nuclei is quite striking; it is, of course, common to all models in which the bulk of the interaction is due to spin (and isobaric spin) independent forces.

The nature of the model is such that it is not expected to be valid for very small or very large A. For low A the statistical assumptions are incorrect. Nevertheless, the formulas which should be valid only for large A apply remarkably well also for light nuclei. For large A the tensor forces appear to become important and invalidate the assumption of spin independence.

CHAPTER 7
Nuclear Models B. Independent Particle Models

7.1. General Features of the Independent Particle or Shell Models

The theory of the isobaric spin and the uniform model do not provide definite wave functions for the nucleus; they only specify certain properties of such wave functions. On the contrary, the independent particle or shell models do give a more or less explicit expression for the whole wave function whence expressions for all the properties of the nucleus in question can be derived.

As was pointed out in Chapter 2, many of the properties of nuclei vary fairly smoothly with the numbers N, Z. Thus the binding energies of nuclei can be represented pretty well by three smooth surfaces for the even-even, even-odd, odd-odd species. This fact was the starting point of the Uniform Model. Closer scrutiny reveals that there are strong discontinuities in some nuclear properties. At certain values of N, Z these discontinuities are particularly marked. These values: 2, 8, 20, 50, 82, 126 are called *magic numbers*. Nuclei with magic N or Z (or both) are called *magic nuclei*. The special stability of the magic nuclei manifests itself in various ways. The nuclei with magic Z or N have larger numbers of stable isotopes (or stable isotones) than their even Z or N neighbors; they are also anomalously abundant indicating a greater stability associated with the magic numbers. The capture cross-sections for neutrons drop sharply at the magic nuclei. All of these features of magic nuclei are attributed to the binding energy discontinuities at the magic numbers. The drop in binding energy may amount to as much as 2 or even 3 mev at the passage of the magic number.

Fluctuations in the properties of nuclei in the neighborhood of the magic numbers can be understood in terms of a "shell" model of the nucleus similar to the highly successful shell model of atoms. On this basis the magic numbers are assumed to indicate the numbers of particles required to provide complete nuclear shells. As in the atomic case, a system consisting of complete shells will show a large separation energy (called ionization potential in the case of atoms). If the number of neutrons or of protons is one larger than can be accommodated in full shells, the last particle will have a small separation energy. According to this picture, magic nuclei play roles among nuclei similar to those of the noble gases among atoms.

The independent particle models for nuclei are similar to the independent particle or Hartree model for atomic electrons. Each nucleon is assumed to move, at least approximately, in an average field set up by the other particles. The individual nucleons are thus effectively independent. The nuclear wave function may then be represented as a product of one-particle wave functions for the individual nucleons except that, in order to satisfy the exclusion principle, these products must be antisymmetrized, i.e. replaced by determinants. The average potential in which the individual nucleons are assumed to move is, or at least resembles, a spherical potential well. The states of the individual particles in such a field can be characterized by five quantum numbers. Three of these are the radial quantum member n (giving the number of zeros of the radial part of the wave function), the orbital angular momentum or azimuthal quantum number l, and the isotopic quantum number τ specifying the type of particle ($\tau = 1$ for neutron, $\tau = -1$ for proton). For the last two quantum numbers one may take the projections of the orbital angular momentum and of the spin in a given direction, l_z and σ. Alternately, one can use as the last two quantum numbers the total angular momentum j of the particle (which is equal to $l + \frac{1}{2}$ or $l - \frac{1}{2}$) and the projection j_z of this quantity in a given direction. The first set of quantum numbers is more adapted for the L-S coupling model; the latter set, for the j-j model. In the first case it is assumed that the total energy is given, in first approximation, by the n and l values of all the particles present; in the second case the j values of these particles are also given. The reason is that the j-j model assumes a large energy-difference between the states $j = l + \frac{1}{2}$ and $j = l - \frac{1}{2}$ of the individual particles. Hence, in the first case the configuration is described by symbols of the type $(1s)^4 (1p)^2$; in the second case, by symbols of the type $(1s_{1/2})^4 (1p_{3/2})^2$. The index $(3/2)$ on the symbol giving the azimuthal quantum number (s for $l = 0$, p, d, f, g, h, i for $l = 1, 2, 3, 4, 5, 6$, respectively) gives the j value of the state.

It was stated before that the individual particle wave function is a determinant. This implies that a single determinant gives the whole wave function of the independent particle models. This is not correct. We have seen that in the L-S model τ, l_z, and σ do not affect the energy substantially, and that the same is true of τ and j_z in the j-j coupling model. As a result, the symbol of the configuration does not contain these quantum numbers. Hence, these quantum numbers can yet be given arbitrary values in the wave functions of the individual particles, and the total wave function may be a sum of determinants formed of individual particle wave functions, the n, l of which are given by the symbol of the configuration, but the τ, l_z and σ of which may be different in each determinant. This applies for the L-S model; in the j-j model the n, l, and j values of the individual particle wave functions are specified by the configuration but τ and j_z can assume any value in each determinant. In order to obtain the coefficients of the various possible determinants of the wave function, one can use the requirements that the total angular momentum J, its projection J_z, the isobaric spin T, and T_ζ have given values. The L-S model assumes, in addition, the validity of the supermultiplet theory. These requirements usually determine the coefficients of the various possible determinants. If not, the corresponding model does not give a unique wave function and one has to resort to calculations based on further assumptions to obtain a wave function for the corresponding state of the nucleus.

7.2. The L-S Coupling Shell Model

In this model the symbol of the configuration gives the radial and azimuthal quantum numbers of the individual particles. The lowest states of a particle in a spherical well are the four 1s states ($n = 1$, $l = 0$, $\tau = \pm 1$, $l_z = 0$, $\sigma = \pm 1$), the next twelve are 1p states ($n = 1$, $l = 1$, $\tau = \pm 1$, $l_z = \pm 1$ or 0, $\sigma = \pm 1$). The 2s and 1d states follow these. Note that the radial or principal quantum number is counted from 1 for all l. Hence the lowest p level is called 1p; the lowest d level, 1d, rather than 2p and 3d as in atomic spectroscopy. The lowest configuration of a nucleus of mass A, for A between 4 and 16, is $(1s)^4 (1p)^{A-4}$. Such a configuration gives rise, in general, to several states with various J and T values. These states are further classified according to their symmetry properties as in the uniform model. If the forces are predominantly ordinary or Majorana forces, the states most nearly symmetric in the positional coordinates (lowest P, P', P'') will lead to the lowest energy values. This is assumed to be true in the simplest version of the shell model which we are considering. Hence, this model uses the assumptions of the supermultiplet theory and is in fact a specialization of that theory which assumes that the wave functions are sums of determinants as described above. It follows, in particular, that the spin angular momentum S is a good quantum number; that it assumes the values 0 and $\frac{1}{2}$ for the low-lying states of even-even and even-odd or odd-even nuclei. The spin is 1 for the $T = 0$ states, and 0 for the $T = 1$ states of odd-odd nuclei with $T_\zeta = 0$ and $S = 1$ or 0 for other odd-odd nuclei.

Except for B^{10} the model gives the parity, the orbital angular momentum, and the order of magnitude of the energy of the excited states of light nuclei. It also explains the difference between favored and unfavored β-transitions. (Chapter 11). This last point constitutes, along with the explanation of the general trend of the energy differences of isobars, the greatest success of the uniform and of the L-S coupling shell models. However, in order to obtain the spin (J value) of the normal states of nuclei, if both L and S are different from zero, one has to obtain the relative positions of the levels which originate from the different orientations of L and S with respect to each other—that is, the levels with J values between $|L - S|$ and $L + S$. These levels all would have the same energy if the forces were entirely independent of the spin. The magnitude and sign of the energy differences between these levels depend on the nature and strength of the spin dependent forces. Very little is known on this last point, and it appears that the type of spin dependent forces such as (5.3), which are suggested by the current meson theories, do not lead to agreement with experiment. Considerable progress has been made, on the other hand, toward the explanation not only of the properties of the normal state but also of the position of the excited states by assuming other types of spin dependent forces. The agreement obtained in this way is often quite striking. At the same time, the calculations with these spin dependent forces also give an indication of the accuracy of the basic picture, i.e. of the wave functions obtained under the neglect of the spin dependent forces. The results are quite favorable: even very strong spin dependent forces yield wave functions in which the wave function of the supermultiplet theory has a probability of about ninety percent. However, this last type of calculation refers principally to nuclei in which a new shell has just been started and there are only a few particles in it. This is the case, for instance, for $A = 18$, for which the $1s$ and $1p$ shells are already completed with 4 and 12 particles, respectively, and there are only two particles outside these shells. There are indications that the situation is much less favorable toward the end of a shell, e.g. for $A = 12$ with the configuration $(1s)^4(1p)^8$. The deviation from the picture of the L-S model appears to be quite strong in this case. The reason for this is not yet understood. A similar remark applies for $A = 14$.

7.3. Comparison of the L-S and j-j Shell Models

It was mentioned before that the L-S model loses validity around $A = 50$, and also that it appears to become increasingly inaccurate toward the ends of the shells. Apparently the spin dependent forces become increasingly important under these conditions. One is reminded that the situation is similar for atomic electrons: the Russell-Saunders coupling is most accurate for light elements and particularly so in the first part of each shell. However, the range of validity of Russell-Saunders coupling for atomic electrons far exceeds the validity of the L-S and supermultiplet models for nuclei. In fact, the opposite extreme —the j-j model—is at present by far the most important nuclear model; it coordinates and explains a variety and wealth of phenomena which exceeds the scope of any other known nuclear model.

The appellations L-S model (and supermultiplet model) express the assumptions that the concepts of orbital and spin angular momenta (and the concept of supermultiplets) have considerable accuracy. Expressed somewhat more mathematically, the orbital and spin angular momenta, (or the triplet of quantum numbers P, P', P'') have with a large probability a single value. It also means that the forces which tend to destroy the validity of these quantum numbers are relatively unimportant and do not cause severe changes in energy. In a similar way, the term j-j coupling shell model expresses the assumption that the j values for individual nucleons are good quantum numbers, that the forces which couple the orbital angular momenta l of individual particles to their spins to form a resultant j are more significant than the forces which couple the individual l together to form a resultant orbital angular momentum L for the whole nucleus. The term j-j coupling should also mean that the forces which tend to destroy the validity of the individual quantum numbers j do not cause too large changes in energy, and that the levels of the same configuration are grouped reasonably closely on the energy scale.

It appears that the L-S model and even the much less specific supermultiplet model are in severe conflict with the j-j model. This is, of course, not true if there is a single particle outside of closed shells, as in this case the individual l is already the L of the whole nucleus and this is, therefore, a good quantum number in both theories. Similarly, j is identical with the total J and is also a good quantum number. The same is true for closed shells or if only a single particle is missing from a closed shell. The L-S models and j-j models are identical also if the open shell is an s shell as the total angular momentum of each particle is necessarily $\frac{1}{2}$ if its orbital angular momentum vanishes. However, even if we have only two particles in the p shell, the wave functions of the two theories are very different. Thus the L-S model postulates for the lowest state of Li^6 the wave function

$$C(x_1 x_2 + y_1 y_2 + z_1 z_2) p(r_1) p(r_2) \delta_{\sigma_1,1} \delta_{\sigma_2,1}(\delta_{\tau_1,1}\delta_{\tau_2,-1} - \delta_{\tau_1,-1}\delta_{\tau_2,1}) \quad (7.1)$$

for the two particles outside the s^4 shell. This is a $^3S_1$ function (with $J_z = 1$) with $T = 0$, which belongs to the supermultiplet (1, 0, 0). $C = 6^{-1/2}$ is a normalization constant and $p(r)$ is the radial part of the wave function for p particles. The state with the same J, J_z, and T_ζ has, according to the j-j shell model, the wave function:

$$(20)^{-1/2}\{3^{1/2}P^{3/2}_{3/2}(r_1, \sigma_1)P^{3/2}_{-1/2}(r_2, \sigma_2) - 2P^{3/2}_{1/2}(r_1, \sigma_1)P^{3/2}_{1/2}(r_2, \sigma_2)$$
$$+ 3^{1/2}P^{3/2}_{-1/2}(r_2, \sigma_1)P^{3/2}_{3/2}(r_2, \sigma_2)\}(\delta_{\tau_1,1}\delta_{\tau_2,-1} - \delta_{\tau_1,-1}\delta_{\tau_2,1}), \quad (7.2)$$

where

$$P^{3/2}_{3/2}(r, \sigma) = 2^{-1/2}(x + iy)p(r)\delta_{\sigma,1},$$
$$P^{3/2}_{1/2}(r, \sigma) = 6^{-1/2}(x + iy)p(r)\delta_{\sigma,-1} - i(2/3)^{1/2}zp(r)\delta_{\sigma,1}, \quad (7.2a)$$
$$P^{3/2}_{-1/2}(r, \sigma) = -i(2/3)^{1/2}zp(r)\delta_{\sigma,-1} + 6^{-1/2}(x - iy)p(r)\delta_{\sigma,1}.$$

The coefficients in these expressions are the so-called coefficients of the vector addition model (Clebsch-Gordan coefficients). The single particle functions (7.2a) have total angular momentum $j = 3/2$; the lower index of the P is the j_z. In the function (7.2) of the j-j coupling model the quantum numbers which are good in both models, i.e. J, J_z, T, T_ζ have the same values 1, 1, 0, 0 as in (7.1). However, the probabilities for the values 1 and 0 for S and L are not unity in (7.2); other values for these quantities ($S = 0$, $L = 1, 2$) have finite probabilities. On the other hand, $j_1 = j_2 = \frac{3}{2}$ for (7.2), i.e. (7.2) represents a $(1p_{3/2})^2$ state, while for the state (7.1) the configurations $(1p_{3/2})(1p_{1/2})$ and even $(1p_{1/2})^2$ have finite probabilities.

It is not known with certainty whether (7.1) or (7.2) is closer to the actual wave function of the normal state of Li^6. Since this is a light nucleus and has only two particles in the $1p$ shell, (7.1) is presumably more accurate. Both (7.1) and (7.2) give reasonable values for the magnetic moment (0.88 and 0.63); the experimental value of this quantity is 0.822. On the other hand, (7.2) gives a much too large value for the quadrupole moment: 7×10^{-27} cm^2. The quadrupole moment would vanish if (7.1) were accurate, but minor deviations from (7.1) suffice to explain the very small observed moment of less than 5×10^{-28} cm^2. Even more convincingly in favor of the L-S model are the values for lifetimes of β-active nuclei in this region. However, a similar comparison for heavier nuclei would almost surely favor the j-j wave function of the type (7.2), and this would be true particularly for nuclei in which each shell is almost filled.

Consequences of the L-S and j-j models differ in many cases less sharply than one would expect. For instance, according to the L-S model, the low-lying states of Li^6 should be ^3S, ^3D for $T = 0$ and ^1S, ^1D for $T = 1$. This gives levels with $J = 1, 1, 2, 3$ for $T = 0$ and $J = 0, 2$ for $T = 1$. Actually, the first six energy levels of Li^6 have these quantum numbers. However, the j-j model, i.e. the configuration $(p_{3/2})^2$, also gives $J = 1, 3$ for $T = 0$ and $J = 0, 2$ for $T = 1$. Furthermore, the $(p_{3/2})(p_{1/2})$ configuration gives $J = 1, 2$ (for both $T = 0$ and $T = 1$) so that the levels $J = 1, 2$, $T = 0$ find a natural interpretation in the j-j model also. A comparison of known states at the end of the p shell, for instance for C^{14}, would give a different result: the L-S model gives $J = 0, 2$ for $T = 1$ (as for Li^6 or He^6); the j-j model, only one low-lying state for $T = 1$, the $(p_{1/2})^2 \, J = 0$ state. Actually, $J = 0$ for the normal state of C^{14} and the first excited state seems to be at 6.1 mev. The j-j model is more nearly in accord with the observations in this case.

7.4. The j-j Coupling Shell Model

As mentioned before, the configuration in this model specifies not only the radial (principal) and orbital angular momentum quantum numbers of the individual particles but also the relative orientation

7.4 · THE j-j COUPLING MODEL

of orbital and spin angular momenta of each particle. It speaks, for instance, of $1d_{5/2}$ and $1d_{3/2}$ levels, rather than simply of a $1d$ level. In the former, spin and orbital angular momenta are parallel; in the latter, anti-parallel. The model assumes, furthermore, that the parallel orientation gives the lower energy and that the energy difference between parallel and anti-parallel orientations is appreciable. It increases with increasing l. The order of the filling of the single particle levels is illustrated in Table 7.1. An important feature of the j-j model is the subdivision of each shell (except of the s shells) into subshells. Thus, the $1p$ shell is subdivided into a $1p_{3/2}$ and a $1p_{1/2}$ subshell; the energy differences between subshells and between shells are quite comparable.

It is customary in the j-j model to give the configuration for protons and neutrons separately, and to separate them with a semicolon. Thus, the lowest configuration of Li^6 would be written as $(1s)^2 1p_{3/2}; (1s)^2 1p_{3/2}$. The lowest configuration for C^{12} would be $(1s)^2(1p_{3/2})^4; (1s)^2(1p_{3/2})^4$. Since j_z can assume only four values in a $p_{3/2}$ subshell, the normal state of C^{12} contains, both for protons and for neutrons, only closed shells. The normal state of B^{12} comes from the configuration $(1s)^2(1p_{3/2})^3$; $(1s)^2(1p_{3/2})^4(1p_{1/2})$. Since the exclusion principle prohibits more than four neutrons in the $1p_{3/2}$ subshell, one neutron must be in the next subshell. This is the $1p_{1/2}$ subshell. One often omits some of the closed shells and writes, for instance, for the lowest configuration of B^{12}, simply $(1p_{3/2})^3; (1p_{3/2})^4(1p_{1/2})$.

This notation gives not only the configuration in the sense in which this term was used before, but also the value of $T_\zeta = \frac{1}{2}(N - Z)$. The symbol of the configuration as used before can be obtained from the new symbol simply by contraction. Thus we would have used in the preceding section the symbol $(1s)^4(1p_{3/2})^7(1p_{1/2})$ for the lowest configuration of B^{12}, and added $T_\zeta = 1$. Conversely, this symbol can be translated into the present notation if the configuration is the lowest for the nucleus in question. In such a case there can be only one incomplete subshell for the protons and one for the neutrons. Hence, the joint symbol can be separated only in one way into a part before the semicolon and a part after the semicolon. This is not the case if the configuration is not the lowest: thus $(1s)^4(1p_{3/2})^7(1p_{1/2})$ for $T_\zeta = 0(C^{12})$ can be separated into $(1s)^2(1p_{3/2})^4; (1s)^2(1p_{3/2})^3(1p_{1/2})$, or into $(1s)^2(1p_{3/2})^3(1p_{1/2}); (1s)^2(1p_{3/2})^4$. In this case either the neutron configuration or the proton configuration contains two incomplete subshells. The reason for adopting a symbol for the configuration of a definite nucleus rather than for a set of isobars as in the preceding section is that the isobaric spin quantum number does not give additional clues for the wave function of those states which have the maximum T_ζ consistent with the configuration. The T of the wave functions with this T_ζ is automatically equal to this T_ζ: certainly no part of such a wave function can have a lower T than its T_ζ; higher T do not occur at all in the configuration. Hence, if dealing with the states of highest T_ζ of the configuration—this is the case as we saw for B^{12}, and is also the rule for heavier nuclei where T_ζ is relatively large—the isobaric spin quantum number is not useful for the determination of the wave function of low-lying levels. The table of successive subshells, i.e. energy levels of individual particles follows. The lowest energy

Table 7.1

$1s_{1/2}(2)$	$1p_{3/2}(6)$	$1d_{5/2}(14)$	$1f_{7/2}(28)$	$2d_{5/2}(56)$	$2f_{7/2}(90)$	$1i_{11/2}(138)$
	$1p_{1/2}(8)$	$2s_{1/2}(16)$	$2p_{3/2}(32)$	$1g_{7/2}(64)$	$1h_{9/2}(100)$	$2g_{9/2}(148)$
		$1d_{3/2}(20)$	$1f_{5/2}(38)$	$3s_{1/2}(66)$	$3p_{3/2}(104)$	$2g_{7/2}(156)$
			$2p_{1/2}(40)$	$1h_{11/2}(78)$	$2f_{5/2}(110)$	$3d_{5/2}(162)$
			$1g_{9/2}(50)$	$2d_{3/2}(82)$	$3p_{1/2}(112)$	$3d_{3/2}(166)$
					$1i_{13/2}(126)$	$4s_{1/2}(168)$
						$1j_{15/2}(184)$

level is in the first column, and in order to obtain the levels in the order of their energy values, one has to read next the second column down; then proceed downward on the third column, and so on. The purpose of the columnar arrangement is to unite all those levels into a column the energy values of which are not sufficiently different to cause a noticeable discontinuity at the transition from one to the next. On the other hand, the energy difference between the last level in a column and the first level in the next one is so large that the nucleus is "magic" (particularly stable) when all the levels of a column are filled. The numbers in the parentheses after the symbol of each subshell indicate the number of neutrons which the subshell in question, and all the subshells of lower energy, can contain. The corresponding numbers for protons are, of course, the same. It follows that the parenthesized numbers at the end of the columns are the magic numbers. The number of neutrons or of protons which a subshell with angular momentum j can accommodate is $2j + 1$, corresponding to the $2j + 1$ values $-j, -j+1, \cdots, j-1, j$ which the quantum number j_z can assume. Hence, each parenthesized number differs by $2j + 1$ from the preceding one.

An important feature of the scheme is the large spin-orbit splitting assumed for the $1g$, $1h$, and $1i$ levels; for these the $j = l + \frac{1}{2}$ and $j = l - \frac{1}{2}$ subshells are in different columns. The order for the levels in Table 7.1 is that obtained for a square well potential. For an oscillator potential ($V \sim r^2$), all the levels in a column except the $1g_{9/2}$, $1h_{11/2}$, $1i_{13/2}$, $1j_{15/2}$ have the same energy; the energy of these levels is the same as that of all levels of the next column. Actually the spin-orbit splitting plays a role also within the levels of a single column. However, the order of the levels within a column does not seem to be entirely fixed; it may change from nucleus to nucleus. Similar phenomena occur also for the electronic shells of atoms.

It is worth noting that all the subshells of one column have the same parity except the $1g_{9/2}$, $1h_{11/2}$, $1i_{13/2}$, $1j_{15/2}$, which have opposite parity to the remaining states of the columns in which they appear.

7.5. Coupling Rules for the *j-j* Model

If we assume a particular order for the one-particle levels, the configuration of the normal state is uniquely given. However, in general, several states of the whole nucleus, with different J values, belong to each configuration. All these states have the same energy in first approximation, so that in order to find the characteristics, in particular the J value, of the normal state, one has to establish an ordering of the states arising from the same configuration. The rules which establish this order, or at least give the J value of the state of lowest energy are called coupling rules.

Originally, these coupling rules were formulated on the basis of empirical evidence. It was postulated that an even number of protons always couple, for the normal state, to $J_p = 0$; likewise an even number of neutrons couple to $J_n = 0$. This already specifies the J of even-even nuclei to be $J = 0$. Similarly, it was postulated that an odd number of protons couple to the J value $J_p = j_p$ of the protons which are in an unfilled subshell—there should be only one such subshell. The same rule is postulated for neutrons if their number is odd. This gives for the J value of even-odd and odd-even nuclei uniquely the j value of the nucleon of which there is an odd number in the nucleus. The above rules do not give a unique answer to the question of the J value of the normal state of odd-odd nuclei—this can be anywhere between $|j_p - j_n|$ and $j_p + j_n$, where j_p and j_n are the j values of the unfilled proton and neutron subshells.

More recently, attempts have been made to provide a theoretical basis for coupling rules. Such attempts necessarily involve assumptions concerning the nuclear interaction responsible for the splitting of the several levels due to a configuration. The assumptions which lead to the above coupling rules are: (1) The interaction is much stronger within the group of protons and within the group of neutrons than *between* protons and neutrons unless these are in the same subshell. (2) The interaction has extremely short range and is attractive. This last assumption makes it unnecessary to specify whether the interaction is of ordinary or Majorana exchange type: if the wave function is antisymmetric with respect to the interchange of the spatial coordinates of two particles it must vanish at their coincidence. Then it will be so small within the assumed range of the forces that the interaction will vanish between particles for which the wave function is antisymmetric in the spatial coordinates, i.e. the interaction will vanish whenever the Majorana interaction is different from the ordinary interaction. Although these assumptions are not fully consonant with our knowledge of nuclear forces from other sources, they do lead to the empirical coupling scheme. Rule (1) is particularly difficult to understand: it leads to grossly inaccurate results if applied to light nuclei. For these, the postulate of the validity of the isobaric spin is in conflict with (1), and provides other coupling rules leading to good agreement with experiment. Although, as has been pointed out before, for heavier nuclei the isobaric spin does not provide coupling rules and is therefore not in conflict with (1), one does not see why postulate (1) should be valid for heavier nuclei if it is not valid for light ones.[1]

Prediction of the quantum numbers of the normal state remains difficult even assuming the above coupling rules. In the first place, the actual order of the one-particle levels (the subshells of Table 7.1) is uncertain. Furthermore, the configuration splitting may lead to a crossing of levels arising from close configurations. Indeed, in order to understand the values of the nuclear spins, and in particular the rather small spins which occur for nuclear ground states of even-odd nuclei, an explicit assumption of such crossing is required. The absence of spins of 11/2 and 13/2 can be explained by assuming that the interaction between the nucleons in a shell increases with increasing l and is proportional to the number of *pairs* of nucleons in the unfilled shell. Thus, for a set of seventy nucleons, the lowest configuration should fill, say, the $3s_{1/2}$ level. However, it may be that the configuration with 2 nucleons in the $1h_{11/2}$ level and with the $3s_{1/2}$ level empty leads to a lower state than the previous configuration because of the increased interaction between the $h_{11/2}$ nucleons. If this is the case, the $3s_{1/2}$ shell will not be filled by seventy nucleons, and the seventy-first nucleon can still be in a $3s_{1/2}$ state. As a result, a set of seventy-one nucleons will give a spin of $\frac{1}{2}$ rather than 11/2.

Magnetic moments predicted by the model for the odd A nuclei are given by the Schmidt lines (see Chapter 3). Magnetic moments of a number of nuclei do lie on these lines but, in general, the moments are

[1] Since these lines were written, M. Redlich (personal communication) has exhibited a case in which the calculation yields, under the assumption of a wide separation of the $j = l + \frac{1}{2}$ and the $j = l - \frac{1}{2}$ levels, a surprisingly small interaction between the protons and the neutrons. He showed that the $J_p = 0$ level, arising from the $(1h_{9/2})^2$ configuration of two protons, lies so much deeper than the other levels arising from the same configuration ($J_p = 2, 4, 6, 8$) that the interaction with a single $2g_{9/2}$ neutron does not introduce appreciable admixtures of the higher J_p states into the wave function of the lowest level of the $(1h_{9/2})^2$; $2g_{9/2}$ configuration. The generality of this result has not been established.

not well described by the model (they are given about equally accurately by several other models). However, the magnetic moments may be used to provide a qualitative check of the assumptions of the model. If the theory were exact so that the magnetic moments were given by the Schmidt lines, then from the spin J of an odd A nucleus, the $j(=J)$ *and* the l value of the last odd nucleon could be determined. We may use the magnetic moments to infer the l value by taking l to be that of the Schmidt line closer to the observed magnetic moment. This l value may then be compared with that which is to be expected on the assumed one-particle level structure, consistent, of course, with the given j. In all but a handful of cases there is agreement with the theory. This result provides powerful support for the level grouping which has been assumed and gives confidence in the determination of the parities of nuclear ground states by means of the model.

7.6. Normal States and Low-Excited States

The values of the magic numbers 50, 82, and 126 provided the incentive and the first justification for the j-j model. However, evidence for the usefulness and validity of this model can now be found in almost every branch of nuclear physics. Perhaps the most impressive and convincing data which support this model derive from investigation of the normal states and low-excited states of nuclei with mass numbers in excess of seventy. Admittedly, the spin and parity assignments of these states are partially based on guesswork, influenced, no doubt, by the shell model which is to be proved. Nevertheless, the possibility of interpreting almost every low level of these nuclei on the basis of the j-j model remains a striking confirmation of the model.

Table 7.2 gives the parities and (for typographical reasons) twice the J value of the low states for odd-even and even-odd nuclei as far as they are known in the region of the Table. The rows refer to Z or N, whichever is odd, the columns to the even one among Z and N. In the left side of the Table the rows refer to N, the columns to Z; in the right side it is the other way around. If more than one symbol is given for a Z-N combination, the lowest refers to the normal state, the one above it to the first excited state, and so on. For the normal states of stable nuclei, J can be measured directly. The parity was obtained from the magnetic moment as explained before: + indicates even parity (even l of the odd proton or neutron),—indicates odd parity (odd l of the nucleon of which there is an odd number in the nucleus). The assignment of J and parity for the excited states and for unstable nuclei is based on the multipolarity of the γ rays emitted by the excited states (Chapter 12) and on the degree of forbiddenness of the β-radiation of unstable nuclei (Chapter 11). Some of the assignments may be open to question but the bulk of them can be expected to prove correct.

TABLE 7.2

	30	32	34	36	38	40	42	44	46	48	50	52	54	56	58	60	62	64	66	68	70	72	74	76	78	80	82
39	$+9 \\ -1$	$-5 \\ -1$	$+9$																								
41	$+9 \\ -1$	$-1 \\ +5 \\ +9$	$+5$	-1							$-1 \\ +9$		$-1 \\ +9$	$-1 \\ +9$													
43		$-3? \\ +7 \\ -1$	$-3 \\ +7 \\ -1$	$+7 \\ -1$	-1					$-1 \\ +9$		$-1 \\ +9$		$-1 \\ +7 \\ +9$													
45		$-1 \\ +7$	$-1 \\ +7$	$-1 \\ +7$											$-1 \\ +7$	$-1 \\ +7$											
47			$+7 \\ -1$	$-1 \\ +7 \\ +9$	$-1 \\ +7 \\ +9$	$-1 \\ +9$	$-1 \\ +9$								$+7 \\ -1$	$+7 \\ -1$	-1	-1									
49			$-1 \\ +9$	$-1 \\ +9$	$-3 \\ -1 \\ +9$	$-1 \\ +9$	$-1 \\ +9$										$+9$	$-1 \\ +9$									
51					$+5$	$+5$															$+5$	$+5 \\ +7$	$+5 \\ +7$	$+7$	$+7$	$+7$	
53					$+5$		$+5$														$+5$	$+5$	$+5$	$+7$	$+7$	$+7$	
55					$+5$	$+5$																		$+5$	$+5 \\ +7$	$+5 \\ +7$	$+7$

In the region between 41 and 49 the $g_{9/2}$ subshell should be filled; the parity and spin ideally should be $+9/2$ throughout this region. Actually, some of the normal states are $p_{1/2}$ states. Explanation of this in terms of pairing energy was given before. The frequent occurrence of $J = 7/2$ with even parity is more difficult to understand. Such a state occurs only for nuclei in which N or Z assume the values 43, 45, or 47. It does not occur for 41 or 49. Naturally the $(g_{9/2})$ and the $(g_{9/2})^9$ configurations do not give a $J = 7/2$ level. However, three, five, or seven $g_{9/2}$ particles do. This suggests that the $J = 7/2$ levels are other members of the $(g_{9/2})^n$ configuration, for $n = 3, 5, 7$, which, according to the coupling rules, should lie above the $J = 9/2$ level. This is the case, as a rule, but not invariably. The most surprising and apparently well established exception is the $J = 5/2$ state of positive parity in $_{41}Se_{34}$.

The occurrence of low-lying states with $J = j - 1$ is not restricted to the $g_{9/2}$ subshell. It is a phenomenon which is rather common in all shells. Thus, the spins of Na^{23}, Ne^{21} are 3/2 instead of 5/2; those of Ti^{47}, Mn^{55} are 5/2 instead of 7/2. The former two nuclei contain three $d_{5/2}$ particles, the latter ones five $f_{7/2}$ particles. The spin of V^{47} with three $f_{7/2}$ particles is 3/2.

While the $p_{1/2}$ and $g_{9/2}$ states compete as long as the odd particle number is below 50, above 50 the competition is between the $g_{7/2}$ and $d_{5/2}$ states. This is quite in agreement with Table 7.1. It is striking, furthermore, how independent the low states are of the number of the particles of which there is an even number in the nucleus. Thus, if there are 45 protons in the nucleus, or 45 neutrons, the lowest state remains invariably a $J = 7/2$ even state; the first-excited state is a $J = \frac{1}{2}$ odd state. This is in full accord with the coupling rules and their explanation: an even number of neutrons (protons) couples to zero total angular momentum, and the interaction of these neutrons (protons) with the protons (neutrons) is so weak that they do not affect the relative position of the states of the latter.

7.7. Magnetic and Quadrupole Moments

The reason for the deviations of the measured magnetic moments from the Schmidt lines (mentioned under Coupling Rules) has been the subject of much discussion. There is general agreement that the intrinsic moments of the nucleons (cf. Chapter 1) will be affected by the proximity of other nucleons. It appears quite likely, however, that this effect is quite small—of the order of a couple of tenths of nuclear magnetons. Since most of the moments lie quite far from the Schmidt lines, one has to assume that the wave function obtained by the simple coupling rules is inaccurate. This view is strengthened by the fact that the calculated magnetic moments agree quite well with the measured ones when one has a more reliable wave function. This is the case for light nuclei, the wave functions of which can be obtained on the basis of the isobaric spin concept. In fact, a study of the magnetic moments of these nuclei indicates that the changes in the intrinsic moment of nucleons tend to shift the magnetic moments beyond the Schmidt lines (cf. H^3, He^3, C^{13}) rather than pull them between these lines.

The modifications of the wave function to bring about agreement between observed and calculated moments do not seem to be large. It is worth recalling that the wave function of a state which is described with 95% probability by the unperturbed state ψ_0 and contains a 5% admixture of ψ_1 is

$$0.975\psi_0 + 0.225\psi_1.$$

If a property such as the magnetic moment has a matrix element connecting ψ_0 with ψ_1, its coefficient will be $2 \times 0.975 \times 0.225 = 0.44$. This shows that a relatively small admixture can radically change some of the properties of stationary states. Let us denote the wave function of the shell model as given by the coupling rules by ψ_0, the operator of the magnetic moment by μ_z. Then if we denote the actual wave function by $\alpha_0\psi_0 + \alpha_1\psi_1$, the total magnetic moment will be

$$(\alpha_0\psi_0 + \alpha_1\psi_1, \mu_z(\alpha_0\psi_0 + \alpha_1\psi_1)). \tag{7.3}$$

It appears that one can obtain reasonable agreement with the observed moment by omitting the $(\psi_1, \mu_z\psi_1)$ term of (7.3) as small. The Rayleigh-Schrödinger perturbation method gives a series expansion for ψ_1 but $(\psi_0, \mu_z\psi_1)$ and $(\psi_1, \mu_z\psi_0)$ is different from zero only for a few of the terms of the series. Their addition to the Schmidt value $(\psi_0, \mu_z\psi_0)$ greatly improves the agreement with the experimental values.

The agreement between observed magnetic moments and those calculated in the way described in the preceding paragraph may not be convincing because almost every modification of the wave function shifts the moments from the Schmidt lines to the region between these lines. However, a similar calculation of the quadrupole moments also greatly improves the agreement with the observed quadrupole moments. There is one exception to this rule: the very large quadrupole moments of nuclei in the rare earth region amounting to several barns (10^{-24} cm^2) cannot be obtained in this way. These quadrupole moments are too large to be accounted for by a single particle with a reasonable wave function. The explanation of these quadrupole moments requires a more drastic modification of the wave function in which, then, the ψ_0 will play only a relatively small role.

7.8. Problems of the *j-j* Model

The general qualitative success of the shell model, at least for not too light nuclei, is most remarkable in view of the simplicity of the

7.8 · PROBLEMS OF THE j-j MODEL

assumptions on which it is based. Several problems remain, however, which it is well to record.

The first stems from the fact that the independent particle model—the representation of the wave function by one or a few simple determinants—can hardly be expected to be accurate. This circumstance together with the great success of the shell model invites a reformulation of the independent particle model in which the determinants are replaced by more general functions. Following the initiative provided by the work of Brueckner and Watson, several promising attempts have been made in this direction. These are formulated in a rather abstract mathematical language so that it is not easy to describe them in terms of the behavior of the wave function. It seems, nevertheless, that the wave function shows a characteristic departure from the wave function of the independent particle model in those parts of the configuration space where two particles are very close together. Furthermore, this departure is the same for all not too highly excited states so that it hardly affects the matrix elements of single particle operators (such as are responsible for electromagnetic or β transitions) between these states.

Second, the origin of the spin-orbit splitting, so characteristic for the model, is unclear. Meson theories do not yield a "vector force," i.e. an interaction of the form:

$$(\boldsymbol{p}_1 - \boldsymbol{p}_2) \times (\boldsymbol{r}_1 - \boldsymbol{r}_2) \cdot (\boldsymbol{s}_1 + \boldsymbol{s}_2) V(r_{12}), \qquad (7.4)$$

which would give the most simple and direct interpretation of the energy differences between the $j = l + \frac{1}{2}$ and the $j = l - \frac{1}{2}$ individual particle levels. Attempts to explain these energy differences as second order effects of the tensor interaction have met only with partial success. The problem of a more detailed explanation of the coupling rules was mentioned before.

The preceding problems concern the foundations of the shell model from first principles. The problems to be enumerated below concern experimental results not accounted for by the model. The best known among these is the failure of the j-j model to distinguish between favored and unfavored β-transitions. Comparison of the L-S and j-j models carried out before showed that the more detailed consideration of the interaction between the particles, which does lead to the explanation of the unfavored nature of most β-transitions, so drastically modifies the wave function that this largely conforms again with the supermultiplet theory. It should be mentioned, however, that no favored β-transitions exist for heavier nuclei (nor can any be expected on the basis of the supermultiplet theory), so that this problem applies only to relatively light nuclei ($A < 50$).

The problem of very large quadrupole moments is taken up again in the next chapter.

If the independent particle picture were accurate, the energy of a nuclear level would be determined, at least in first approximation, by the configuration from which it arises. That this is not the case can be seen most clearly by enumerating all the energy levels of the whole nucleus which arise from the lowest configuration in the middle of a shell. Consider first a small A at which the isobaric spin is still a good quantum number—let us say $A = 8$. The lowest configuration is $(1s)^4(1p_{3/2})^4$; it gives the following levels: $J = 4, 2, 2, 0$ for $T = 0$, $J = 3, 2, 1$ for $T = 1$ and $J = 0$ for $T = 2$. Among the $T = 1$ states the $J = 2$ is lowest. Its position is known from the energy difference between the normal states of Be^8 and Li^8: it lies 17.0 mev above the normal state. The position of the $J = 0$, $T = 2$ state cannot be obtained from available information but can be estimated to lie about 12 mev above the $T = 1$, $J = 2$ state. Thus the $(1s)^4(1p_{3/2})^4$ configuration is spread out over 29 mev. It certainly does not form a closely spaced group of levels as one would expect if the independent particle picture were valid. Naturally, $A = 8$ is a low mass number, and one will not expect the j-j model to be very accurate. However, a similar situation prevails in general as long as neutrons and protons are in the same shell.

The j-j model underestimates the energy differences between isobars, and the calculations which improve on this situation by considering the forces in more detail again modify the wave functions so drastically that they conform largely with the supermultiplet theory.

In some of the more recent modifications of the j-j shell model, the attempt to explain the relative position of the levels by means of nuclear potentials similar to (5.6) is entirely abandoned. Instead, a largely phenomenological approach is followed which is based on the two assumptions that the wave function is correctly given by the j-j model, and that the interaction takes place between pairs of nuclei.

Let us consider, for instance, the $d_{5/2}$ shell. The $(d_{5/2})^2$ configuration gives six states with total angular momenta $J = 0, 1, 2, 3, 4, 5$. The $J = 0, 2, 4$ are isotopic spin triplets; the $J = 1, 3, 5$ are isotopic spin singlets, but this fact does not enter the following considerations. The next step is, ideally, the determination of the energy values of these states. This must be done experimentally. However, the six energy values of the two-particle system suffice to obtain the position of all energy levels of the rest of the $d_{5/2}$ shell from F^{19} to Si^{28}. In order to obtain, for instance, the energy levels of the $(d_{5/2})^4$ configuration, one decomposes the wave function $\psi(1, 2, 3, 4)$ of a level of this configuration, as given by the shell model, in the following way:

$$\psi(1, 2, 3, 4) = \sum_{J=0}^{5} \sum_{m=-J}^{J} u_{Jm}(1, 2) f_{Jm}(3, 4). \tag{7.5}$$

The u_{Jm} are the wave functions of the two-nucleon problem mentioned above. Every antisymmetrized product $D_\mu^{5/2}(1) D_{\mu'}^{5/2}(2) - D_{\mu'}^{5/2}(1) D_\mu^{5/2}(2)$ can be expressed in terms of these u_{Jm}: the four-particle wave function $\psi(1, 2, 3, 4)$ of the j-j model, as function of the coordinates of the particles 1 and 2, is a linear combination of such antisymmetrized products. The $f_{Jm}(3, 4)$ in (7.5) are simply the expansion coefficients of $\psi(1, 2, 3, 4)$ in terms of the u_{Jm}; they depend, naturally, on the coordinates of the particles 3 and 4.

The probability w_{Jm} that the particles 1 and 2 are in the state u_{Jm} is

$$w_{Jm} = \int |f_{Jm}(3, 4)|^2 \, d3 \, d4, \tag{7.5a}$$

where the integration includes summation over spin and isotopic-spin variables. Hence, the interaction energy between particles 1 and 2 is, for the state $\psi(1, 2, 3, 4)$:

$$\mathcal{E}_{12} = \sum_{J=0}^{5} \sum_m w_{Jm} E_J \tag{7.5b}$$

where E_J is the energy of the two-particle state with total angular momentum J, as obtained experimentally. The interaction energy between all other pairs is equally large so that the total energy in the state described by ψ becomes

$$E = \frac{4 \times 3}{2} \mathcal{E}_{12}. \tag{7.5c}$$

The factor $4 \times 3/2$ is the number of pairs of $d_{5/2}$ particles.

If the position of the six states of the $(d_{5/2})^2$ configuration is not known, and the energy values of all levels of a configuration rarely are, the energy values of some of the levels of higher configurations must be used to determine the E_J. One has, at any rate, six constants available to fit all the levels of all the configurations $(d_{5/2})^n$, up to $(d_{5/2})^{12}$. The number of constants is $2j + 1$ if the angular momentum of the particles in the shell being filled is j.

The preceding calculation is based on the assumptions: (a) that the j-j model does give accurate wave functions in terms of individual particle wave functions and that these do not change substantially while the shell is being completed, and (b) that the nuclear forces are too complicated to permit the calculation of the interaction even if the wave function is known. It is assumed that the total interaction is the sum of interactions between pairs of particles.

In many cases (7.5c) gives surprisingly accurate values. There are substantial deviations from (7.5c) in other cases. This had to be expected since interconfigurational interaction is surely not always negligible.

If it is strong, the actual wave function will differ substantially from that given by the shell model. It will neither be known, nor could it be expanded in the form (7.5) if it were known. In spite of this, the success of the formal treatment of the interaction indicates that the j-j model is accurate in some sense, at least, in a large number of cases.

CHAPTER 8

Nuclear Models III Many Particle Models

8.1. The α Particle Model

Correlations between motions of individual nucleons which may exist in addition to those required by the exclusion principle are entirely neglected in the independent particle model; they are described only in a global fashion in the uniform model. The α particle model is based on the assumption that these correlations are, in all nuclei, very similar to those in the α particle. It is suggested that a group of two neutrons and two protons forms a tight configuration in the nucleus and moves within the nucleus as a unit. On this picture the nucleus is considered as similar to a molecule, the individual units of which are α particles. It is not essential to presume that the α particles have a permanent identity. They may form, dissolve, and form again. If the time during which the α particle structure is maintained is long compared with the intervening period, the properties of the nucleus may be calculated with good approximation on the basis of the molecular picture.

The binding energies of nuclei composed of an integral number of α particles are well accounted for by the α particle model if, but only if, one disregards the increasing electrostatic energy. The binding energy of the α particles themselves accounts for a large part of the total binding energy. The remaining binding energy arises through the interactions between the α particles. It appears that the binding energy per α particle bond in these nuclei is approximately constant (except for the case of Be^8 which does not form a bound state). This feature contributes strongly to the attractiveness of the model.

In general, however, the model has met only moderate success. It is difficult to reconcile the form of the model with the data on the α-α scattering and the absence of binding in the Be^8 system. The nuclei which are not composed of α particle units are not easily incorporated into the model.

8.2. Collective Model

The very large quadrupole moments in the rare earth region suggest that some nuclei are strongly non-spherical. Thus, the quadrupole moments of Lu^{175} and Ta^{181} are around 6×10^{-24} cm^2, and in view of the nuclear densities given in Chapter 3, this corresponds to a rotational ellipsoid with major and minor half-axes of 7.7×10^{-13} and 6.2×10^{-13} cm. This is then the order of magnitude of the deformation of the whole nuclear matter which contains seventy-one or seventy-three protons. It is quite unlikely that such large deformations could be explained by the non-spherical nature of the wave function of a single proton, or even of several protons; the modification of the nuclear wave function probably extends to all protons or at least to a large fraction of them.

The large deviations from the spherical shape are, perhaps, not so surprising since the energy of nuclear matter inside the nucleus does not depend on the form of its surface: the nucleons which are further from the surface than the range of nuclear forces are in the same surroundings, no matter what the shape of the surface. The only obvious difference in the energy content of spherical and non-spherical nuclei is that more nucleons are close to the surface in an ellipsoidal nucleus than in a spherical one, or that the surface energy of the former is greater than in the latter. However, the difference between the surface area of an ellipsoid and that of a sphere of equal volume is only of the order of the square of the difference of major and minor axes. A more detailed calculation shows that the excess surface of the ellipsoid postulated for the two strongly deformed nuclei is only of the order $(8/45)(7.7-6.2)^2 6.7^2$ or about one percent of the total surface. With the surface energy as given by (2.1) this amounts to 3 mev. The increased surface energy can be compensated by the increased ease with which the aspherical h orbits can be fitted into an ellipsoidal surface, i.e. by the effect of the centrifugal forces of nucleons with rather high orbital angular momenta, and by tensor forces. The latter favor, on the whole, a prolate rather than an oblate shape, and their general effect manifests itself in the preponderance of the positive quadrupole moments. The centrifugal effect favors an oblate shape for less than half-filled shells and a prolate shape if the shell is more than half filled. The observed quadrupole moments on the whole follow this trend (cf. Fig. 3.2).

The relatively easy deformability of nuclear matter is in marked contrast to the definitely spherical shape of the electronic cloud of atoms. This is subjected to the attractive force emanating from a center which is spherically symmetric and gives the atomic shape considerable rigidity. The easily deformable nuclear matter will adjust to some degree even to the wave function of a single particle which may be concentrated close to a particular plane. The nucleons with high angular momenta have such strongly aspherical wave functions.

Further elaboration of these ideas led to the concept of a characteristic *shape* of the nucleus. The constant density of the nuclei implies that for every configuration of the nucleons which occurs with appreciable probability, there is a more or less well-defined surface which surrounds all nucleons and the inside of which is filled rather uniformly by the nucleons. For light nuclei this surface has spherical shape. The large quadrupole moments indicate, however, that the shape may be different for heavy nuclei, and the simplest assumption is that it is a rotational ellipsoid. The orientation of the rotational ellipsoid will not be the same for all configurations, and the ratio of the axes of the ellipsoid may also show a certain amount of variation. The first circumstance is associated with the rotation of the nuclear ellipsoid as a whole; the second, with vibrations of the ellipsoid. Naturally these concepts have to be modified to some extent because of the diffuse nature of the nuclear surface (Section 3.2), but this modification is rather trivial.

The quantities which specify the orientation and the shape of the ellipsoid are called collective variables. Together with internal coordinates which specify the positions of the nucleons with respect to the ellipsoid, these collective coordinates specify the positions of all nucleons in space, and the wave function is considered to be a function of both the internal and the collective coordinates. The dependence of the wave function on the collective coordinates shows a great resemblance to the dependence of the wave function of a diatomic molecule on the coordinates of the two nuclei, and the whole theory is rather similar to the theory of diatomic molecules. In particular, the energy is independent of the orientation of the rotational ellipsoid in space, just as it is independent of the orientation of the internuclear axis of diatomic molecules. However, the energy does depend on the rapidity of the variation of the nuclear wave function, as a function of the angles describing the ellipsoid's orientation in space. This corresponds to rotational kinetic energy of the ellipsoid. This kinetic energy is zero if the wave function is independent of the orientation ($J = 0$), but assumes the value:

$$\frac{\hbar^2}{2I} J(J + 1) \tag{8.1}$$

if the wave function depends as the spherical harmonic $P^J(\theta, \phi)$ on the polar angles θ, ϕ of the principal axis of the ellipsoid; $J\hbar$ is then the total angular momentum. I plays the role of a moment of inertia. It is not, however, the total moment of inertia of the ellipsoid but only of that part of it which can be set into rotation. This is not the case for the closed shells, the wave functions of which are spherically symmetric so that they cannot be set into rotation. Hence, the I is smaller than the whole moment of inertia of the nuclear ellipsoid. The range of values which the total angular momentum J can assume also shows a great similarity to the range of values of the analogous quantity for diatomic molecules. If the nucleons rotate in the ellipsoid and if their angular momentum about the principal axis of the ellipsoid is $K\hbar$, the range of the quantum number J becomes $K, K + 1, K + 2, \cdots$. This is always the case for nuclei with odd mass numbers where K is half integer. In even-even nuclei $K = 0$ is the rule, and J ranges, just as for symmetric diatomic molecules, over the numbers $0, 2, 4, \cdots$.

The energy does depend on the collective coordinate which describes the nuclear deformation, and also on the rapidity of the variation of the nuclear wave function, as function of this collective coordinate. The former dependence corresponds to the elastic energy connected with the deformation from the most stable shape; the latter, to the kinetic energy of the oscillations around this shape. The simplest assumptions lead to a vibrational spectrum but this will not be discussed in detail.

The experimental discovery of the rotational spectra is the outstanding success of the collective model. There are, in particular, two regions in which a score of rotational levels has been observed, the energies of which are given with an accuracy of about one percent by (8.1). In the first of these regions the number of neutrons is between 90 and 114; in the second region the number of protons exceeds 88. The upper boundary of the second region is not known; there are no nuclei with $Z > 100$. The same two regions are characterized by large quadrupole moments ranging up to 10 barns (for Er^{167} with 99 neutrons and 68 protons).

Outside of the two regions the accuracy of (8.1) is not very good. In particular, the ratio of the excitation energies of the $J = 4$ and $J = 2$ states in even-even nuclei, for which (8.1) gives the value 3.3—and which is 3.3 in the two regions—drops to about 2.3. The quadrupole moment is also smaller outside the two regions and the constant in (8.1) larger, indicating a more nearly spherical nucleus. The coupling conditions outside the regions of the preceding paragraph are evidently less in accordance with the picture of a definite nuclear shape. The transition probabilities between the various rotational states also attest to the correctness of the interpretation of these states; they can be calculated with a reasonable accuracy from the picture and the wave function described above. There is, on the other hand, a serious discrepancy between the deformation as derived from the magnitude of the quadrupole moments or the transition probabilities on the one hand, and the magnitude of the deformation obtained on the original, straightforward interpretation of the constant in (8.1). The latter is too small by a factor of about five.

8.3. Comparison of the *j-j* and the Collective Models

Although even light nuclei show certain signs of a collective behavior, the regions in which the *L-S* and the collective models are valid are on the whole so far removed from each other that it is not necessary to trace the boundary between the realms of these models. On the other hand, the *j-j* model and the collective model claim validity in contiguous regions, so that it would be desirable to describe the relation between these models in detail.

On the whole, the *j-j* model is most accurate in the vicinity of magic

8.3 · j-j COUPLING AND COLLECTIVE MODELS

numbers, both below and above these. Also in these regions the consequences of this model are least ambiguous: if there are only a few particles outside the closed shells or if only a few particles are missing from otherwise closed shells, the configuration at least is given rather definitely. The collective model, on the other hand, is most accurate far from magic nuclei, and in the two regions of particular validity, Z or N is between 90 and 114—about 10 units removed from the magic numbers 82 and 126. The transition between the regions of validity of the two models is rather abrupt and there are only a few nuclei in the twilight region.

The concepts with which the j-j and the collective models operate are so different that it is not possible to give the difference between them as unambiguously as the difference between L-S and j-j wave functions could be given in the preceding chapter. The wave functions of the j-j model with the usual coupling rules cannot be correct because their quadrupole moments are much smaller than the typical large quadrupole moments. These range up to 10×10^{-24} cm^2, while the quadrupole moments of the j-j model do not exceed about 0.3×10^{-24} cm^2 even for the heaviest nuclei. Even the configurational interaction described in Section 7.7 does not more than double this figure.

The situation is even worse in nuclei with an odd number of neutrons and an even number of protons, in which the protons should couple to a state with angular momentum $J_p = 0$ and give no quadrupole moment in first approximation. The large quadrupole moments must be due to several protons in orbits elongated in the same direction, but the ordinary coupling rules do not yield such orbits.

Actually, in the regions in which the collective model is particularly useful, so many orbits compete with each other that there is little reason to believe in the validity of the simple coupling rules. These were established for the situation in which only a single proton subshell and a single neutron subshell were open—that is, not completely filled. It is conceivable that a wave function which conforms with the collective model could be obtained if the complex situation of several open subshells could be handled mathematically. It is, on the other hand, also possible that the wave functions of the individual particles undergo considerable modification in the regions of the collective model. The relatively large quadrupole moment of 0.8×10^{-24} cm^2 of In is an indication in this direction. The proton configuration is unique in this case and contains only one unfilled orbit: a $g_{9/2}$ orbit. The quadrupole moment associated with this orbit is only 0.2×10^{-24} cm^2 and it has this value only if the neutrons couple to $J_n = 0$. Otherwise, the quadrupole moment is even smaller. Configuration interaction almost doubles the quadrupole moment but it falls, even then, short of the experimental value. This example and some others, therefore, indicate that not only the coupling rules but also the orbits of the j-j model are inaccurate in the regions of the collective model's validity.

A great weakness of the collective model is its great flexibility. As a matter of fact, in spite of many promising attempts to specify properties of the wave function demanded by the collective model, there is no unanimity in this regard. When the discrepancy between the equilibrium shapes, as obtained from quadrupole moments and transition probabilities on the one hand and the positions of the rotational levels (8.1) on the other hand, became known, it was at once possible to modify the theory in such a way that the discrepancy was no longer apparent. Another weakness of the theory is that the limits of its validity are not apparent from the theory itself. This is, of course, hardly possible as long as its postulates are not formulated in the language of quantum mechanics. Naturally, the large quadrupole moments are a fact, and so are the large transition probabilities between the rotational states. These phenomena clearly point to a deformed nucleus. However, the question whether the dynamics of the deformed nucleus is as simple as the collective model implies remains to be seen.

The greatest success of the collective model is that it predicted the rotational states and the large transition probabilities between these states both by γ-emission and Coulomb excitation, even before these states were found experimentally. In addition, it has been applied lately to the phenomenon of fission.

CHAPTER 9

Nuclear Reactions A. Close Collisions

Nuclear reactions are collision phenomena and almost all methods developed for the treatment and understanding of collision phenomena have been used to interpret nuclear reactions. Many of these methods have, indeed, been developed for this purpose. Chapter 5 quotes some results for two-body problems—only these can be treated accurately with present-day techniques. If more than two particles are involved in the reaction, i.e. in all cases except proton-proton and proton-neutron scattering, one must be content either with approximate results based on models similar to those used in Chapters 6, 7, and 8, or with general and accurate formulas which, however, contain several adjustable constants obtainable only experimentally or by using approximate calculations. We begin with an example for the second method.

9.1. The Collision Matrix

A collision between nuclei a and X may in general give rise to a variety of reaction products:

$$a + X \to \begin{cases} a + X & (s) \\ a + X^* & (t) \\ b + Y & (u) \\ c + Z & (w) \end{cases}, \qquad (9.1)$$

where X^* is an excited state of the nuclear system X. The first reaction is called elastic scattering, the second inelastic scattering, the remainder exchange reactions. We consider only reactions in which two reaction products are formed. Each pair of products may be indicated by an index (s, t, etc.). The possibility of capture of a by X with subsequent γ-emission is not considered in the analysis below; we take up such processes in relation to the considerations of resonance phenomena.

The nuclear system C consisting of all the nucleons in X and a (or Y and b, etc.) may be described in a many-dimensional configuration space. We distinguish two regions in this space—(1) the *internal region* within which the separations of all nucleons are of the order of nuclear dimensions, (2) the external region where at least the separations between some nucleons are greater than nuclear dimensions. For reasons which arise later, we imagine a many-dimensional surface S dividing the two regions; S is taken wholly in the external region, but everywhere near the internal region.

For a given energy E of the incident nuclei, only a limited set of reaction products are energetically possible, and the energies available for the relative motion of any pair of products are definite. As a consequence, the wave function for the system at energy E will be different from zero in the external region only in certain "channels" of the configuration space which describe the system C as a pair of nuclei $a + X$, $b + Y$, etc. A particular state of motion of well separated products t may be indicated by $R_t(\mathbf{r}_t)\chi_t$, where χ_t represents the state of the nuclei defined by the index t, and $R_t(\mathbf{r}_t)$ describes the relative motion of the pair t of nuclei. The specifically nuclear forces have no influence on the function $R_t(\mathbf{r}_t)$. These functions will depend for a given E only on the presence or absence of the electrostatic field between the pair t, the relative angular momentum of the pair, and boundary conditions.

For simplicity we assume that the spins of all nuclei involved in the reaction are zero. The removal of this assumption complicates the argument but adds nothing of importance in principle. We consider, to begin with, a collision in which the two nuclei of the pair s converge with relative orbital angular momentum quantum number l, i.e. the incoming part of the wave function in the s channel is proportional to $r_s^{-1} P_l(\mathbf{\Omega}_s) I_{ls}(r_s) \chi_s$, where $r_s^{-1} P_l(\mathbf{\Omega}_s) I_{ls}(r_s)$ represents an incoming spherical wave of orbital angular momentum l. The direction of the inter-nuclear line is $\mathbf{\Omega}_s$, its length r_s. The whole wave function for the system in the "open" channels—the channels which represent energetically possible reaction products—will consist, in the external region, of this incident wave together with emergent waves of the same angular momentum:

$$\Phi_l = (M_s/\hbar)^{1/2} P_l(\mathbf{\Omega}_s) r_s^{-1} I_{ls}(r_s) \chi_s \qquad (9.2)$$
$$- \sum_t S_{st}^l (M_t/\hbar)^{1/2} P_l(\mathbf{\Omega}_t) r_t^{-1} E_{lt}(r_t) \chi_t,$$

where $P_l(\mathbf{\Omega}_t) r_t^{-1} E_{lt}(r_t)$ represents an emergent wave of orbital angular momentum l in the channel t. The I_{lt} and E_{lt} can be chosen as conjugate complex functions which satisfy at every r_t the relation:

$$I_{lt} E'_{lt} - E_{lt} I'_{lt} = I_{lt} I'^*_{lt} - I^*_{lt} I'_{lt} = 2i, \qquad (9.2a)$$

in which the primed quantities I'_{lt}, E'_{lt} denote the derivatives of I_{lt}, E_{lt} with respect to r_t, the internucleon separation. M_t is the reduced mass of the pair t and $(M_t/\hbar)^{1/2} I_{lt}$ and $(M_t/\hbar)^{1/2} E_{lt}$ are waves with unit inward and outward current per unit solid angle. The numbers S_{st}^l describe the amplitudes of the emergent waves for the given incident wave; they are dependent, of course, on the nature of the collision, i.e.

on the character of the colliding pair s, the product nuclei t, the angular momentum l, and also on the total energy E of the system. The quantities S^l_{st} completely characterize the collision properties of the system. If only elastic scattering is possible, $S^l_{ss} = \exp 2i\delta_l$, where δ_l is the so-called phase shift. It is convenient to consider the set of quantities S^l_{st} as an entity—the collision matrix for the system.

If no interaction were present between the particles of the pair s, the wave function for a system in which these particles stream toward each other in the z direction and with unit flux per unit area would be simply $(M_s/\hbar k_s)^{1/2} \chi_s \exp\{ik_s r_s (\mathbf{e}_z \cdot \mathbf{\Omega}_s)\}$. In order to compare this expression with (9.2), one decomposes it into spherical waves; each such spherical wave represents a definite angular momentum:

$$(M_s/\hbar k_s)^{1/2} \exp\{ik_s r_s(\mathbf{e}_z \cdot \mathbf{\Omega}_s)\}\chi_s$$
$$= \sum_l \tfrac{1}{2}(2l+1)i^{l+1}(M_s/\hbar)^{1/2}P_l(\mathbf{\Omega}_s)k_s^{-1}r_s^{-1}(I_l - E_l)\chi_s. \tag{9.3}$$

This is a purely mathematical identity; $\mathbf{e}_z$ is the unit vector in the direction of motion of the incident particle in the pair s; the $P_l(\mathbf{\Omega}_s)$ are the well known spherical harmonics; $P_l(\mathbf{e}_z) = 1$. The wave number k_s is the relative momentum of the colliding (or separating) pair s, divided by $\hbar$. Hence $(\hbar^2/2M_s)k_s^2 = E - \mathcal{E}_s$ where $\mathcal{E}_s$ is the inner energy of the particles of the pair s. The k_t are defined in a similar way in terms of the inner energies $\mathcal{E}_t$ of the pairs t; the difference $\mathcal{E}_s - \mathcal{E}_t$ is the energy release (Q value) of the reaction $s \to t$.

If one superposes the wave functions (9.2) which take the interaction into account with amplitudes $\tfrac{1}{2}(2l+1)i^{l+1}k_s^{-1}$, a wave function

$$\Phi = \sum_l \tfrac{1}{2}(2l+1)i^{l+1}k_s^{-1}\Phi_l \tag{9.2b}$$

is obtained, the incident waves in which are identical with those of (9.3). Φ gives, for the external region, the wave function for the collision of the particles s with unit flux, taking interaction into account. The difference between (9.2b) and (9.3) gives the products of the collision:

$$\Phi' = \sum_l \sum_t (S^l_{st} - \delta_{st})\tfrac{1}{2}(2l+1)i^{l+1}(M_t/\hbar)^{1/2}P_l(\mathbf{\Omega}_t)k_t^{-1}r_t^{-1}E_{lt}(r_t)\chi_t. \tag{9.2c}$$

The coefficient of χ_t describes the production of the pair t; the coefficient of χ_s, the elastic scattering. If there is no interaction between the particles, the difference (9.2c) is zero, that is the collision matrices S^l are all unit matrices in this case. The outgoing waves (9.2c) can be observed only for large separation r_t of the pairs; for large r_t the asymptotic form of E_{lt} is:

$$E_{lt}(r_t) = I_{lt}(r_t)^* \sim k_t^{-1/2}i^{-l}e^{ik_t r_t}. \tag{9.3a}$$

The reaction and scattering cross-sections are all implicit in (9.2c).

Since the integral of the square of $P_l(\Omega)$ over all directions is $4\pi/(2l+1)$, the total cross-sections become:

$$\sigma_{st} = \frac{\pi}{k_s^2} \sum_l (2l+1) |S_{st}^l|^2 \quad \text{(reaction)}, \quad (9.4)$$

$$\sigma_{ss} = \frac{\pi}{k_s^2} \sum_l (2l+1) |S_{ss}^l - 1|^2 \quad \text{(scattering)}. \quad (9.4a)$$

The last equation applies only if one of the pair s is uncharged, i.e. is a neutron. Otherwise, because of the long-range nature of the electrostatic interaction, the scattering extends to such high l that the last expression ceases to be useful and the formulas must be modified to take the electrostatic scattering into account more directly.

Certain properties of the collision matrix elements follow from general considerations. From the assumption of only two-particle breakup in the reaction, the number of pairs of incident particles per second is equal to the total number of emergent pairs per second. From this it follows that $\sum_t |S_{st}^l|^2 = 1$, and one can conclude also that $\sum_t S_{st}^l S_{s't}^{l*} = 0$, for $s \neq s'$. An investigation of the properties of the system under time reversal shows further that $S_{st} = S_{ts}$: the collision matrix is unitary and symmetric. The symmetry relation leads to the so-called *reciprocity theorem*:

$$k_\alpha^2 \sigma_{\alpha\beta} = k_\beta^2 \sigma_{\beta\alpha}, \quad (9.4b)$$

which can be derived also from the thermodynamic principle of detailed balance.

Some properties, or more properly, some general limitations on the properties of collision processes may be derived from the results so far obtained, which, it will be noted, do not involve the nature of the interactions between the colliding nuclei. For a particular pair of colliding systems, upper limits on the reaction and scattering cross-sections can be established. One can also obtain the general character of the expected angular distribution. For example, for incident neutrons, if the wave length of the incident beam $\lambdabar = \lambda/2\pi = 1/k$ is very much larger than the nuclear diameter, only the $l = 0$ wave (s wave) will be effective in the collision. The impact parameters for higher orbital angular momenta are large, and the probability of effective collision is therefore small. In this case, all the angular distributions are spherically symmetric. The maximum scattering cross-section occurs for $U_{ss}^0 = -1$ and is $\sigma_{ss} = 4\pi/k_s^2$. The maximum total reaction cross-section $\sum_{t \neq s} \sigma_{st}$ is π/k_s^2 (for $S_{ss}^0 = 0$). The associated scattering cross-section is also π/k_s^2. The angular distributions for scattering and reaction are isotropic in the center of mass coordinate system.

For the *total* cross-section, (9.3) and (9.3a) give:

$$\sum_t \sigma_{st} + \sigma_{ss}$$

$$= \frac{\pi}{k_s^2} \sum_l (2l+1)\{\sum_t |S_{st}^l|^2 + |S_{ss}^l|^2 - S_{ss}^l - S_{ss}^{l*} + 1\} \qquad (9.3b)$$

$$= \frac{\pi}{k_s^2} \sum_l (2l+1)\{2 - S_{ss}^l - S_{ss}^{l*}\}.$$

The sums in this equation exclude $t = s$, and the last line follows from $\sum |S_{st}^l|^2 = 1$ as extended over all t. The total cross-section for the collision of the pair s depends only on the first power of the S_{ss}^l.

9.2. Qualitative Discussion of Resonance Phenomena

Under certain conditions the cross-section σ_{st} for the reaction $a + X = Y + b$ rises, passes through a more or less sharp maximum, and falls to normal values as the energy of the incident particle a passes through certain energy intervals. The energy of the whole system (in the center of mass coordinate system) at the maximum cross-section is referred to as the resonance energy E_λ of the reaction; the width in energy of the maximum in the cross-section function is called the width Γ_λ of the resonance level. The index λ specifies the resonance level. Examples for typical resonance behavior of the cross-section are given in Fig. 9.1.

This general behavior of the cross-section may be understood on the basis of the picture given already in Section 4.2. We assume that there exists a quasi-stationary state with average energy E_λ of the compound nuclear system C composed of the nucleons of the particles a and X. Since this state is quasi-stationary, its energy is not precisely defined; we denote the uncertainty in the energy of the state by Γ_λ. The reaction is presumed to proceed in two steps: 1. The incident particles make a transition to the quasi-stationary state of the compound nucleus. 2. This unstable state decays with the formation of the reaction products. The process is similar to that involved in the resonance scattering of radiation by atoms. On the basis of this picture one may speak of the cross-section σ_{sC} for the formation of the compound nucleus. This may be shown to be (we assume that only the $l = 0$ state is effective):

$$\sigma_{sC} = \frac{\pi}{k_s^2} \frac{\Gamma_{\lambda s}\Gamma_\lambda}{(E_\lambda - E)^2 + \frac{1}{4}\Gamma_\lambda^2}, \qquad (9.6)$$

where $\Gamma_{\lambda s}$ measures the probability of transition from the initial state of the incident s particles to the state λ of the compound nucleus. It is proportional to the square of a suitably formulated matrix element between these states. The behavior of the compound nucleus after

76 9 · CLOSE COLLISIONS

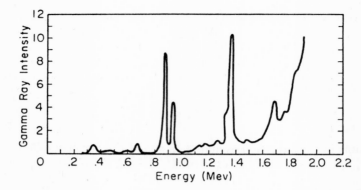

Figure 9.1a. Cross-Section of the (p, γ) Reaction in Fluorine. The abscissa is the energy of the incident protons in mev; the ordinate, the γ-ray yield in arbitrary units. From E. J. Bernet, R. G. Herb, and D. B. Parkinson, *Phys. Rev.* 54, 398 (1938), one of the early observations of the resonance phenomenon.

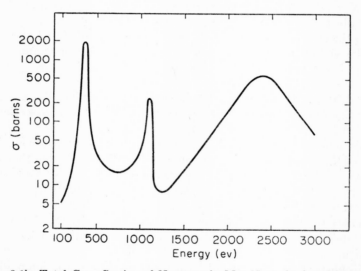

Figure 9.1b. Total Cross-Section of Neutrons in Mn. Note the logarithmic scale for the cross-section which is given in barns. From L. M. Bollinger, D. A. Dahlberg, R. R. Palmer, G. E. Thomas, *Phys. Rev.* 100, 126 (1955).

formation depends on the properties of the state of the compound nucleus only. This compound nucleus breaks up with the formation of the pair t or the formation of other pairs. We write for the probability of the breakup of the compound nucleus into the t pair: $\Gamma_{\lambda t}/\Gamma_\lambda$, where $\Gamma_{\lambda t}$ is the partial width of the λ level for the emission of the pair t. Thus the cross-section for the formation of the t particles becomes:

$$\sigma_{st} = \frac{\pi}{k_s^2} \frac{\Gamma_{\lambda s}\Gamma_\lambda}{(E_\lambda - E)^2 + \tfrac{1}{4}\Gamma_\lambda^2} \frac{\Gamma_{\lambda t}}{\Gamma_\lambda} = \frac{\pi}{k_s^2} \frac{\Gamma_{\lambda s}\Gamma_{\lambda t}}{(E_\lambda - E)^2 + \tfrac{1}{4}\Gamma_\lambda^2}. \quad (9.6a)$$

$\Gamma_{\lambda t}$ has a role similar to $\Gamma_{\lambda s}$: it is proportional to the square of a matrix element between the state of the compound nucleus and the final state of the t particles. Clearly we have $\sum_t \Gamma_{\lambda t}/\Gamma_\lambda = 1$, since the intermediate state of the compound nucleus must break up in some way. This formalism permits also the description of the capture of a by X with subsequent emission of radiation. The compound system may radiate rather than break up by particle emission, and we may assign a partial width for radiation $\Gamma_{\lambda r}$ to this process. $\Gamma_{\lambda r}/\hbar$ measures the probability-per-second that the excited state of the compound nucleus, formed in the collision, goes over to a lower state of the system by emission of radiation.

9.3. Derivation of the Resonance Formula

The preceding result is valid only to the extent to which the notion of a long-lived compound nuclear state is valid. The general condition for this is that the width Γ_λ be much smaller than the separation between adjacent levels.

It is difficult to judge the validity of the assumptions made above for actual nuclear processes. This can be done, however, on the basis of a mathematical derivation of (9.6a) using quantum mechanical theory. We have seen that the cross-sections of collision processes are given by the collision matrix. The elements of this matrix can be obtained from a knowledge of the stationary states of the system throughout the whole configuration space. Indeed what is required for the determination of the collision matrix elements are only the properties of the stationary state wave functions on the parts of the surface S (which separates internal and external regions of configuration space), where this surface is traversed by the open channels. If the wave function is known at this surface, the requirement of continuity across the surface completely determines the wave function in the open channels outside S, and hence determines the collision matrix elements. The quantities which characterize the properties of the state at the surface of the channel t may be taken as: 1.) The coefficients v_t^l of $P_l(\Omega_t) r_t^{-1} \chi_t$ in the expansion of a wave function Φ on the surface S, and 2.) the similar quantities d_t^l in the expansion of the radial derivative $\partial \Phi/\partial r_t$ of the same wave function. These are related by a set of linear equations, valid for all channels s,

$$M_s^{-1/2} v_s^l = \sum_t R_{st}^l M_t^{-1/2} d_t^l. \tag{9.7}$$

M_s, M_t are again the reduced masses of the pairs s, t. The quantities R_{st}^l which enter this relation may be considered as elements of a matrix R^l. This matrix provides a generalization to reaction processes of the concept of the logarithmic derivative for potential scattering. Naturally, one has a different matrix R^l for each value of l. It can be shown that the matrix elements of R^l, as functions of the energy E of the system, can be expanded into a series of the form

$$R^l_{st} = \sum_\lambda \frac{\gamma_{\lambda s}\gamma_{\lambda t}}{E_\lambda - E}, \qquad (9.8)$$

where the $\gamma_{\lambda s}$, $\gamma_{\lambda t}$, E_λ are real, energy-independent constants. They are different for different l. It follows from (9.8) that R^l is real and symmetric. The E_λ are the eigenvalues of an Hermitean boundary-value problem $HX_\lambda = E_\lambda X_\lambda$ for the internal region specified by the Hamiltonian operator H of the system, together with certain boundary conditions on S. The transition strengths $\gamma_{\lambda s}$ give, *on the surface S*, the normalized characteristic function X_λ in terms of the channel functions χ_t:

$$X_\lambda = \sum_t \frac{(2M_t)^{1/2}}{\hbar} \gamma_{\lambda t} \left(\frac{2l+1}{4\pi}\right)^{1/2} P_l(\Omega_t) r_t^{-1} \chi_t. \qquad (9.8a)$$

This equation is valid only on the surface S. Since the expression after $\gamma_{\lambda t}$ in (9.8a) is normalized, this equation shows that $(2M_t/\hbar^2)\,\gamma_{\lambda t}^2$ can be thought of as the probability that the stationary state X_λ consists, on the surface S, of the pair t.

The matrix R^l is determined at all energies by the parameters E_λ, $\gamma_{\lambda s}$. The S^l matrix, and hence the cross-sections, can be obtained from the R^l matrix by postulating the validity of (9.7) for the Φ^l of (9.2). This leads to:

$$S^l = (I_l - I'_l R^l)(E_l - E'_l R^l)^{-1} = (I_l - I'_l R^l)(I_l^* - I'^*_l R^l)^{-1}. \qquad (9.9)$$

In this, I_l and $E_l = I_l^*$ are diagonal matrices, the diagonal elements of which are the values of $I_{ls}(r_s)$ and $E_{ls}(r_s)$ for the r_s which correspond to the intersection of the channel s with the surface S. Similarly, the diagonal elements of the diagonal matrices I'_l and $E'_l = I'^*_l$ are the radial derivatives of these functions at the same r_s. One obtains by means of (9.4), (9.8), and (9.9) the same formula (9.6) for the reaction cross-section which the more visualizable derivation at (9.6) gave, provided one can approximate R^l by a single term in the expansion (9.8). If this is the case, the E_λ play, apart from a minor correction discussed below, the roles of resonance energies. The partial widths are related to the $\gamma_{\lambda s}$ through

$$\Gamma_{\lambda s} = \frac{2\gamma_{\lambda s}^2}{|I_{ls}|^2}. \qquad (9.10)$$

For given $\gamma_{\lambda s}$ the partial width $\Gamma_{\lambda s}$ and hence also the probability of the reaction product s decreases with increasing I_{ls}. If the incident particle of the pair s is a neutron, the I_{ls} are spherical waves of free particles. In particular, for $l = 0$

$$I_{0s}(r_s) = k_s^{-1/2} \exp(-ik_s r_s). \tag{9.11}$$

In this case $\Gamma_{\lambda s} = 2k_{\lambda s}\gamma_{\lambda s}^2$: the neutron width is proportional to the wave number $k_{\lambda s}$ and hence to the velocity of the neutron. From (9.6) the cross-section for a neutron capture reaction is, at very low neutron energy, inversely proportional to the neutron velocity—a result which follows from the above considerations even if one does not approximate (9.8) by a single term, and has therefore general validity.

If both particles of the pair s are charged, $I_{ls}(r_s)$ is a Coulomb wave function, i.e. the wave function of a particle moving in an inverse square field of force. It is much more complicated than (9.11). It is customary in such a case to write:

$$|I_{ls}|^2 = 1/k_s p, \tag{9.12}$$

and call p the penetration factor. It expresses the fact that the density of the particles must be p^{-1} times higher at S for charged particles than for neutrons in order to give the same current at a very large distance. Accurate expressions and useful approximations for p were given in particular by Breit and his collaborators. A crude but in many cases useful approximation for $l = 0$ is:

$$p = \exp\left\{-\frac{\pi ZZ'e^2}{\hbar}\left(\frac{2M_s}{E - \mathcal{E}_s}\right)^{1/2} + \frac{4}{\hbar}(2M_s ZZ'e^2 a_s)^{1/2}\right\}. \tag{9.12a}$$

In this, $ZZ'e^2$ is the product of the charges on the two nuclei of the pair s, their kinetic energy at large separation is $E - \mathcal{E}_s$, and a_s is the value of the radius r_s at which the channel s intersects the surface S. The formula is valid only if the first term in the bracket is considerably greater than the second. It gives a more quantitative measure for the decreased probability of charged particle reactions than the qualitative considerations of Chapter 4. For the case of neutrons, i.e. in the absence of electrostatic interaction, $p^{-1} = 1$ for $l = 0$ and is given for higher l as function of ka_s in Fig. 9.2.

The expression (9.10) for $\Gamma_{\lambda s}$ contains two factors: the first of these, $\gamma_{\lambda s}^2$ depends only on the properties of the internal region, while $|I_{ls}|^{-2} = k_s p$ depends on the behavior of the wave function in the channels, that is in the external region. For this reason $\gamma_{\lambda s}^2$ is often called the reduced width of the level λ for disintegration into the pair s, because the actual width $\Gamma_{\lambda s}$ would have this value for $p = 1$, $k_s = 1$. Division of the $\Gamma_{\lambda s}$ by $k_s p$ eliminates the factors characteristic for the external region.

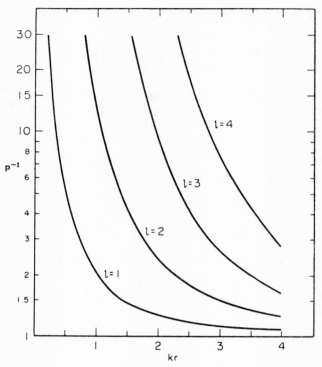

Figure 9.2. Reciprocal Penetration Factors for Neutrons. The reciprocal penetration factors are given for neutrons with various angular momenta $l\hbar$ as function of the product kr of the wave number of the neutron and the radius of the nucleus. The penetration factor is 1 for $l = 0$.

9.4. Dependence of the Parameters on the Size of the Internal Region

Although the R matrix may always be characterized by parameters E_λ, $\gamma_{\lambda s}$, some care must be exercised in associating these quantities with the observed resonance energies and widths. The parameters E_λ, $\gamma_{\lambda s}$ depend on the position of the surface S surrounding the internal region, while those for the energy of the maximum cross-section and the level widths do not. Hence, the expressions for the latter involve, in addition to the E_λ, $\gamma_{\lambda s}$, other S dependent quantities (the I_{ls} and I'_{ls}) to cancel out the S dependence of the E_λ, $\gamma_{\lambda s}$. This is apparent, for instance, in (9.9) the left side of which is, naturally, independent of S while both $\gamma_{\lambda s}$ and I_{ls} depend on S. Similarly, the energy for the maximum cross-section, the E_λ of (9.6), is smaller than the E_λ of (9.8) by the real part of $\sum_t \gamma_{\gamma t}^2 I'_{lt}/I_{lt}$. In general, the replacement of (9.8) by a single term will be the more accurate the smaller the internal region is, consistent with the requirement that the specifically nuclear interaction be negligible for points of the configuration space which lie outside S. This suggests laying the surface S so that the channels intersect it close to the nuclear radius; the variables r_s, r_t to be inserted into I_t, I'_t in (9.9) or (9.10) then become equal to the sum of the radii of the two nuclei which constitute the pair of the channel. At the same time one can derive at least a crude estimate for the $\gamma_{\lambda s}$ which corresponds to this position of S if s corresponds to the emission of a single proton or neutron:

$$\gamma_{\lambda s}^2 \sim D/\pi K \ , \tag{9.13}$$

in which D is the average level spacing (i.e. the average distance of successive E_λ) and K the highest wave number of the nucleons in the nucleus. The experimental value for $\gamma_{\lambda s}^2/D$ is about 10^{-13} cm; the derivation of (9.13) will be given later. If the level spacing D and the $\gamma_{\lambda s}$ for processes other than neutron or proton emission are known, the above formulas suffice to obtain estimates for average reaction cross-sections and similar quantities.

For instance, one obtains for the average cross-section of the (n, γ) process:

$$\bar{\sigma}_{n\gamma} \sim \frac{1800 \Gamma_r}{\Gamma_r E^{1/2} + 4.4 \times 10^{-4} DE}. \tag{9.14}$$

The orbital angular momentum of the neutron was assumed to be zero for (9.14); the corresponding processes are the most important ones if the energy E of the neutrons, which is measured in electron volts for (9.14), is below a few kev. Γ_r is the radiation width; it is usually of the order of 0.1 ev. Since (9.13) is only approximate, (9.14) is also only an approximation to the average neutron absorption cross-section. It is quoted here because it shows that the rather abstract formulas of the resonance theory can be reduced to numerical statements.

9.5. Radioactivity

Radioactivity is not properly a problem of collision theory. In spite of this it can be treated by a slight modification of the preceding formalism.

A radioactive state contains only outgoing waves. At the same time the amplitude of the wave function decreases in time at every point of space, reflecting the radioactive decay of the original nucleus. The wave function increases as a function of distance: the particles which are very far away from the source are very numerous because they originate at a time when the source was very much stronger than it is now. Such a wave function is not in accord with the principles of orthodox quantum mechanics; nevertheless, it gives a vivid picture of the process of radioactivity and is the one commonly used for the description thereof.

Let us assume that the wave emerging from the channel s is given by $\alpha_s(M_s/\hbar)^{1/2} P_l(\mathbf{\Omega}_s) r_s^{-1} E_{ls}(r_s)$; it then carries a total current of $4\pi |\alpha_s|^2/(2l+1)$. The v_s on the left side of (9.7) then becomes just $\alpha_s (M_s/\hbar)^{1/2} I_{ls}^*$, the d_t becomes $\alpha_t (M_t/\hbar)^{1/2} I_{lt}'^*$ (we avoid the use of E_{ls} in order to avoid confusion with the energy E). Substituting again a single term for the sum in (9.8), one obtains from (9.7):

$$\alpha_s I_{ls}^* = \frac{\gamma_{\lambda s}}{E_\lambda - E} \sum_t \gamma_{\lambda t} \alpha_t I_{lt}'^*. \tag{9.15}$$

This shows that the $\alpha_s I_{ls}^*$ are proportional to the $\gamma_{\lambda s}$, i.e. the current in the channel s is proportional to:

$$|\alpha_s|^2 = \text{const} \frac{\gamma_{\lambda s}^2}{|I_{ls}|^2} = \text{const } \Gamma_{\lambda s}. \tag{9.16}$$

This confirms the interpretation of the $\Gamma_{\lambda s}$ given at (9.6). The constant of (9.16) corresponds to the fact that the source strength was left arbitrary above; it cannot be determined from (9.15) because these are linear and homogeneous in the α_s. However, the condition that (9.15) has a non-vanishing solution can be obtained by multiplying it with $\gamma_{\lambda s} I_{ls}'^*/I_{ls}^*$ and summing it over all channels s. The sum then drops out on both sides and one obtains for the energy of the radioactive state:

$$E - E_\lambda = -\sum_s \gamma_{\lambda s}^2 I_{ls}'^*/I_{ls}^*. \tag{9.17}$$

This shows that the energy of the radioactive state is not real. Its imaginary part is, because of (9.2a):

$$\frac{E - E^*}{2i} = -\sum_s \frac{\gamma_{\lambda s}^2}{|I_{ls}|^2} = -\tfrac{1}{2} \sum_s \Gamma_{\lambda s} = -\tfrac{1}{2}\Gamma_\lambda. \tag{9.18}$$

Hence, the amplitude of the radioactive state decays everywhere as $\exp(-\tfrac{1}{2}\Gamma_\lambda t/\hbar)$, and the decay constant of its intensity is $\Gamma_\lambda/\hbar$ as we had to expect. The real part of (9.17) is the energy shift discussed in Section 9.4. This discussion of the radioactive decay confirms the qualitative picture of resonance levels developed in section 9.2, as well as the interpretation of the p of Eq. (9.12) as a penetration factor.

9.6. The Clouded Crystal Ball Model

If the nuclear wave functions corresponded to complete chaos as Bohr's original ideas and the powder models (Chapter 6) postulated, the $\gamma_{\lambda s}^2$ averaged over several resonances λ would not be expected to show maxima and minima as functions of the energy of the resonance.

9.6 · THE OPTICAL MODEL

These models which are also called strong coupling models lead one to expect that the cross-sections, after being averaged over several resonances, are smooth functions of the energy. The expression (9.14), which was derived under the assumption (9.13) of the strong coupling model indeed shows such a behavior. On the other hand, the success of the individual particle models leads one to expect that the effect of the target nucleus on the incident nucleon can be represented, at least approximately, by an average potential, or as a region in which the index of refraction of the incident nucleon's wave function is different from 1. This picture leads one to expect considerable fluctuations in the cross-sections. These correspond to the interference maxima and minima in the scattering of light by a crystal ball the size of which is comparable to the wave length of the incident light. Fluctuations in the cross-section which correspond to this picture were indeed found, and it is natural to describe these fluctuations by representing the effect of the target nucleus on the incident particle by a potential well.

The analogy between crystal ball and nucleus is incomplete because the amount of light that enters a non-absorbing crystal ball is equal to the amount that leaves it. On the other hand, the particle which enters the nucleus may react with it so that a different particle may leave the nucleus and the intensity of the emerging wave which corresponds to the original particle alone is smaller than that of the incident beam. This circumstance will appear as an absorption in a theory which concerns itself only with the wave function of the incident particle. In order to account for this absorption, it must be assumed that the crystal ball is not entirely transparent—that it has the capacity to absorb, not only to refract. This leads to the concept of the "clouded" crystal ball which can be described as a region with a complex refractive index, or its quantum mechanical analogue—a region in which the potential is complex. The simplest form of the model assumes a potential:

$$V = V_0(r)(1 + i\xi), \qquad (9.20)$$

where $V_0(r)$ has a fixed magnitude $-V_0$ in the inside of the target nucleus, and vanishes outside of it; ξ is a constant. This picture can be generalized by establishing a smooth transition between the value $-V_0$ and 0 of $V_0(r)$.

The clouded crystal ball model differs from standard theory in two regards. First, by replacing the target nucleus by a potential, it disregards the structure of this nucleus. As a result it has no means for describing the changes in the target nucleus, but gives an expression only for the wave function of the incident particle, that is for elastic scattering. The other processes including inelastic scattering are not distinguished but lumped together as absorption, or rather formation of the compound nucleus with subsequent disintegration thereof. Second, by disregarding the intricate quantum mechanical nature of the interaction, it fails to account for the resonance structure of scattering and absorption processes and gives only *average* values for the cross sections. However, these average values are the ones which are, as a rule, of principal interest, and for these the model presents a very good picture. The value of ξ in (9.20) which gives best agreement for low-energy processes is about 0.05 while the depth V_0 has the same magnitude which accounts for the binding energies of the stable orbits of the shell models. This is about 45 mev for a radius of $1.25 \times 10^{-13} A^{1/3}$ cm. A ξ of about 0.3 would largely smooth out the maxima and minima of the average absorption and scattering and lead to the type of results that were expected on the basis of the strong coupling theories, that is, the original ideas of Bohr or the uniform models.

9.7. The Intermediate Coupling or Giant Resonance Model

The clouded crystal ball model gives a fair description of the average cross-sections and perhaps also of the actual cross-sections in the energy region in which the resonances are so broad that they overlap and their individuality is lost. It remains desirable, however, to give a more detailed description of the nuclear reactions than this model provides, which accounts for the line structure of the cross-sections and in which the average cross-section is obtained as an average over the rapidly fluctuating line structure. It is also desirable to distinguish between the various nuclear reactions which the clouded crystal ball model lumps together and follows only to the point of the "formation of a compound nucleus". Such a more detailed theory can also dispense with the concept of the "absorption" of the incident particle—a concept which has no direct foundation in quantum theory.

A more detailed theory of nuclear reactions can be based on an analysis of the changes which the interaction between the incident nucleon and the individual nucleons of the target nucleus cause in the approximate wave function, in which only the average effect of the target nucleus is taken into account. This latter wave function is obtained if the target nucleus is replaced by a (real) potential, the potential $V^*(r)$ obtained by averaging over the positions of the nucleons of the target nucleus. This is the crystal ball model which, as mentioned before, leads only to scattering but to no nuclear reaction processes. It will give a scattering cross-section which shows, as function of the energy, maxima and minima; these result from the diffraction of the incident beam in the crystal ball. Furthermore, if the potential assumed is the real part of (9.20)—that is, the potential which also accounts for the individual particle levels known from the shell theory (Chapter 7)—the position of the maxima will coincide with the maxima of the *average*

9.7 · GIANT RESONANCE MODEL

cross-section. This follows at once from the fact that the clouded crystal ball model reproduces these maxima and that the imaginary part has little influence on the positions of these, so that the position of the maxima will be the same if one sets $\xi = 0$ in (9.20). The crystal ball model which we are considering now is, naturally, even more inaccurate than the clouded crystal ball model because it accounts in no way for the occurrence of nuclear reactions, while the clouded crystal ball model accounts for it at least in a global fashion. However, the intermediate coupling model also modifies the crystal ball model and does this not by the assumption of an absorption, but by considering the effect of the replacement of the average potential by the sum of the potentials of the individual nucleons of the target nucleus.

The potential $V^s(r)$ will have resonance levels F_k in the sense of Section 9.3; the associated internal states will be denoted by X_k. This represents a state in which the target nucleus is in its lowest state and the incident nucleon has the excess of F_k over this energy; the scattering of this nucleon in the potential V^s has a maximum for this excess energy. The separation between adjacent resonance energies F_k is of the order of 10 to 20 mev—as was mentioned before, this spacing corresponds to the spacing of the maxima of the average cross-section. The maxima are very broad; their reduced widths γ_k^2 are very large.

The system has many other stationary states φ_μ; in these the target nucleus is in one of its excited states and the energy of the incident particle is the excess of the energy of the state φ_μ over the energy of the excited state of the target nucleus. For all of these states φ_μ, the reduced width for elastic scattering is zero because the state of the target nucleus cannot change under the assumption of the crystal ball model. The reduced widths for elastic scattering are shown in Fig. 9.3a for all states: the states with a large reduced width are states X_k, those with zero reduced widths are the φ_μ.

The complete interaction potential will differ, of course, from $V^s(r)$. The internal eigenstates of the actual system, ψ_λ, (energy E_λ), may be represented as linear combinations of the single particle eigenstates:

$$\psi_\lambda = \sum_\mu \varphi_\mu C_{\mu\lambda} + \sum_k X_k b_{k\lambda}. \qquad (9.21)$$

The range of energy Δ of the single particle states which are appreciably mixed to form the nuclear state ψ_λ furnishes a rough estimate of the degree of failure of the single particle model. In the strong coupling model the range Δ is presumed to be large compared with ΔF, the separation of single particle resonance scattering states. Weak coupling, on the other hand, implies that only one single particle state is of importance in the nuclear state ψ_λ, i.e. Δ is small compared with the average separation D between neighboring levels. Neither the very strong nor the very weak coupling theories are capable of providing a description of experimental observations.

The actual condition in nuclei appears to be intermediate between those implied by the strong and weak coupling models. The assumption that the range Δ is large compared with D but small compared with ΔF provides the basis for a qualitative explanation of the form of the

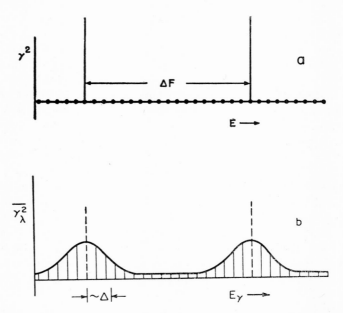

Figure 9.3. Reduced Widths in the Single Particle and Intermediate Coupling Models. The upper part of the figure illustrates the reduced width for scattering in the pure single particle model. The points along the energy axis represent schematically the energy levels in the single particle picture. The reduced width, indicated by the lengths of the vertical lines, is zero for most levels; for a few of them it is very large. In the states corresponding to these, the target nucleus is in the normal state, the incident particle in a high orbit. The lower part of the figure shows the reduced widths in the intermediate coupling model (schematic). The very large widths of some of the levels of the independent particle picture are spread out over several adjacent levels.

$\bar{\sigma}(E, A)$ surface for neutron cross-sections. This is the basic assumption of the "intermediate coupling" model. In this model ψ_λ contains an appreciable contribution from at most one X_k state: that one which is nearest in energy to E_λ:

$$\psi_\lambda = \sum_\mu \varphi_\mu C_{\mu\lambda} + X_k b_{k\lambda}, \qquad |E_\lambda - F_k| < \Delta F/2. \qquad (9.21a)$$

Under these circumstances the reduced width (elastic scattering) for the state ψ_λ is derived entirely from the intermixture in ψ_λ of the single

9.7 · GIANT RESONANCE MODEL

particle state X_k, since for all the other single particle states this reduced width is zero. Thus

$$\gamma_{\lambda s}^2 = |b_{k\lambda}|^2 \gamma_k^2, \tag{9.22}$$

where γ_k^2 is the reduced width of the level F_k. It follows from the completeness of the states ψ_λ that

$$\sum_\lambda \gamma_{\lambda s}^2 = \gamma_k^2 \sum_\lambda |b_{k\lambda}|^2 \cong \gamma_k^2. \tag{9.23}$$

The sum has to be extended only over those states λ which have an appreciable contribution from the state X_k—that is, the group of states around F_k. Thus the effect of the coupling is to spread the original width γ_k^2 over a set of levels in the neighborhood of the energy F_k. The range of energy over which the levels have appreciable reduced widths for elastic scattering is given roughly by Δ. The energy dependence of the *average* of the reduced widths for scattering (averaged over several levels) as expected from this model is shown in Fig. 9.3.b The actual reduced widths will of course fluctuate considerably about these averages.

The order of magnitude of γ_k^2 can be obtained on the basis of the assumption that the single particle wave function has, at the surface of the nucleus, about the same amplitude as inside the nucleus. This gives, by means of the normalization condition of the wave function, $(M/\hbar^2) a\gamma_k^2 = 1$ where a is, as before, the nuclear radius. If the reduced width γ_k^2 is about evenly distributed among all the levels which lie between two levels F_k of the independent particle picture, the average reduced width of the actual levels will be, on account of (9.23), $\gamma_k^2 D/\Delta F$. The spacing ΔF of the single particle levels is about $\hbar^2 K\pi/Ma$ where K is the wave number of the single particle. Hence, on this picture

$$\gamma_{\lambda s}^2 \sim \frac{D}{\Delta F} \gamma_k^2 \sim \frac{D}{\hbar^2 K\pi/Ma} \frac{\hbar^2}{Ma} = \frac{D}{\pi K}. \tag{9.24}$$

This corresponds to the Bethe-Weisskopf estimate (9.13). In the giant resonance model the width γ_k^2 of the single particle level F_k is not evenly distributed among the actual levels and the levels which are close to the broad levels F_k of the single particle model have a substantially greater reduced width than that given by (9.24); the reduced widths of those levels which are far from the F_k are well below (9.24).

The important quantity for the determination of the average value of the scattering elements of the collision matrix is the average value of the scattering reduced width which is represented in the intermediate coupling model in Fig. 9.3b. Thus the average cross-sections can be obtained from the model once the F_k are located and a choice of Δ is made. The F_k are located through the single particle potential $V^*(r)$. This potential is similar to the clouded ball model's $V_0(r)$, while the quantity Δ may be shown to correspond to $V_0\xi$. With this identification of parameters the two models give the same average cross-sections.

Rough estimates of $V^s(r)$ and of Δ may be made on the assumption of a definite law of interaction between nucleons. With simple two-body forces of the character discussed in Section 5.1, one obtains a $V^s(r)$ which is similar to that obtained empirically on the basis of the clouded ball model. However, the calculation of Δ is quite intricate and has not yet been accomplished entirely satisfactorily. The value of Δ obtained by a naive straight-forward calculation corresponds to a strong coupling model. This *may* be an indication that the interactions between nucleons within heavy nuclei are rather different from those obtained in the study of two-nucleon systems.

CHAPTER 10

Nuclear Reactions B. Surface Reactions

The considerations of the last section are valid under all circumstances; they are useful only if R can be simplified sufficiently so that S can be calculated by means of (9.9). That this is not always the case is indicated by our having omitted, in all applications, all but one of the infinitely many terms of (9.8). This is a useful approximation only if the Γ_λ are smaller than the spacing between successive E_λ. Furthermore, the decomposition of the total wave function into spherical waves corresponding to different angular momenta $l\hbar$ gives a similar decomposition only for the total cross-section. In the expression for the differential cross-section, cross terms between waves of various angular momenta appear; the interference of the partial waves of (9.2) cancels only if the cross-section is integrated over all directions. Hence the spherical

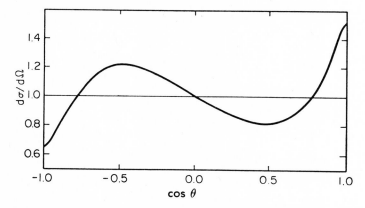

Figure 10.1. Angular Distribution of Scattering (schematic). The figure illustrates the large effect of a small scattering amplitude with a high angular momentum. The non spherically symmetric part (an f-wave) contributes only one per-cent to the total scattering.

wave decomposition will lead to particularly awkward expressions if many angular momenta contribute to the scattered wave. This is most likely to be the case in the forward and backward directions. Fig. 10.1 illustrates this by giving the angular distribution of a wave with $S^0 = i$ (to give one-half of the maximum amount of spherically symmetric scattering) and $S^3 = \exp(-i\pi/50)$. This last part of the scattering contributes hardly more than one per cent to the total cross-section.

It seems reasonable, under the conditions, to attempt applying Born's collision theory to obtain the angular distribution in the forward direction. This theory is always relatively easy to apply. It cannot be expected to reproduce the resonance character of the cross-section, but can be relied upon to give a good value for the average cross-section as long as the partial cross-sections are all small as compared with π/k_s^2. Butler first recognized that the complicated angular distribution of the reaction products points to such a situation for many (d, p) and (d, n) reactions. The following interpretation of the angular distribution of these reactions is due to him, although obtained by him on the basis of another type of mathematical analysis. As before, we disregard the spin of the deuteron as well as that of the emerging proton and neutron: even though they do affect the result of the calculation, they do not influence the underlying principles.

10.1. Angular Distribution in Stripping Reactions

Let us denote the deuteron wave function by $d(\mathbf{r}_p - \mathbf{r}_n)$, where $\mathbf{r}_p$ and $\mathbf{r}_n$ denote the positions of the proton and neutron, respectively. The incident deuterons have the direction of $\mathbf{k}_i$; the wave function of the incident beam is then

$$d(\mathbf{r}_p - \mathbf{r}_n) \exp\left(\tfrac{1}{2}i\mathbf{k}_i \cdot (\mathbf{r}_p + \mathbf{r}_n)\right). \tag{10.1}$$

We substitute a spherically symmetric potential field $V_p(r_p) + V_n(r_n)$ for the target nucleus. This approximation has no major influence either on the calculation or on the result. We calculate the probability of neutron capture into a state $u(\mathbf{r}_n)$ of angular momentum l, with the proton traveling in the direction $\mathbf{k}_e$. The principal interest centers at the relative probabilities of the various directions of $\mathbf{k}_e$, i.e. the angular distribution of the protons. There is, naturally, a similar question concerning the angular distribution of the neutrons resulting from the (d, n) process. Since this can be calculated in the same way as the angular distribution of the protons, we can confine attention to the latter question.

The absolute value of the wave number of the emerging proton is given by the energy Q of the reaction:

$$(\hbar^2/4M)k_i^2 + Q = (\hbar^2/2M)k_e^2, \tag{10.2}$$

in which the proton and deuteron masses are denoted by M and $2M$. According to Born's collision theory the probability of the proton emerging within unit solid angle in the direction $\mathbf{k}_e$ is proportional to the absolute square of:

$$\iint d\mathbf{r}_p \, d\mathbf{r}_n u(\mathbf{r}_n) e^{-i\mathbf{k}_e \cdot \mathbf{r}_p}(V_n(r_n) + V_p(r_p)) \, d(\mathbf{r}_p - \mathbf{r}_n) e^{i\mathbf{k}_i \cdot (\mathbf{r}_p + \mathbf{r}_n)/2}, \tag{10.3}$$

where $d\mathbf{r}_p$ indicates integration over the components of $\mathbf{r}_p$ and $d\mathbf{r}_n$

has a similar significance. There are $2l + 1$ integrals of the form (10.3) if the angular momentum of the bound state u is l, corresponding to the possible orientations of this angular momentum. Hence, the total angular distribution is given by the sum of the absolute squares of $2l + 1$ integrals of the form (10.3).

It is possible to bring the two terms of (10.3) into somewhat simpler forms. Let us write, first:

$$u(\mathbf{r}_n) = (2\pi)^{-3/2} \int U(\mathbf{k}) e^{i\mathbf{k} \cdot \mathbf{r}_n} \, d\mathbf{k}. \tag{10.4}$$

Then

$$V(r_n)u(\mathbf{r}_n) = \left[\frac{\hbar^2}{2M} \Delta + E_n\right] u(\mathbf{r}_n) = \frac{1}{(2\pi)^{3/2}} \int U(\mathbf{k}) \left[E_n - \frac{\hbar^2 k^2}{2M}\right] e^{i\mathbf{k} \cdot \mathbf{r}_n} \, d\mathbf{k},$$

where E_n is the binding energy of the neutron in the potential. It is a negative quantity. $U(\mathbf{k})$ is the wave function of the bound neutron state in momentum space; it has the same angular dependence as u. We can now write for the term of (10.3) which contains $V_n(r_n)$:

$$(2\pi)^{-3/2} \cdot \int d\mathbf{r}_p \int d\mathbf{r}_n \int d\mathbf{k} U(\mathbf{k}) \left(E_n - \frac{\hbar^2 k^2}{2M}\right) d(\mathbf{r}_p - \mathbf{r}_n)$$

$$\cdot \exp i(\mathbf{k} + \tfrac{1}{2}\mathbf{k}_i) \cdot (\mathbf{r}_n - \mathbf{r}_p) \exp i(\mathbf{k}_i - \mathbf{k}_e + \mathbf{k}) \cdot \mathbf{r}_p$$

$$= \iint d\mathbf{r}_p \, d\mathbf{k} U(\mathbf{k}) \left(E_n - \frac{\hbar^2 k^2}{2M}\right) D(\mathbf{k} + \tfrac{1}{2}\mathbf{k}_i) e^{i(\mathbf{k}_i - \mathbf{k}_e + \mathbf{k}) \cdot \mathbf{r}_p}$$

$$= (2\pi)^3 U(\mathbf{k}_e - \mathbf{k}_i) \left(E_n - \frac{\hbar^2 (\mathbf{k}_e - \mathbf{k}_i)^2}{2M}\right) D(\mathbf{k}_e - \tfrac{1}{2}\mathbf{k}_i). \tag{10.5}$$

$D(\mathbf{k})$ is the deuteron wave function in momentum space; since the wave function is, in a very good approximation, spherically symmetric, $D(\mathbf{k}) = D(k)$ depends only on the absolute value of its argument. The last line follows by Fourier's theorem.

As mentioned before, there are $2l + 1$ integrals similar to (10.3) corresponding to the possible orientations of the neutron orbit u. The U in the corresponding $2l + 1$ expressions (10.5) will contain the same dependence $U_r(|\mathbf{k}_e - \mathbf{k}_i|)$ on the absolute value of its argument $\mathbf{k}_e - \mathbf{k}_i$; the dependence on the direction of $\mathbf{k}_e - \mathbf{k}_i$ will be given by the $2l + 1$ spherical harmonics P_{lm}, with $m = -l, -l+1, \cdots, l-1, l$. Omitting the $V_p(r_p)$ term in (10.3), the angular dependence of the emitted protons is given by:

$$J(\mathbf{k}_e) = |U_r(|\mathbf{k}_e - \mathbf{k}_i|)D(|\mathbf{k}_e - \tfrac{1}{2}\mathbf{k}_i|)|^2 [(\hbar^2/2M)(\mathbf{k}_e - \mathbf{k}_i)^2 - E_n]^2. \tag{10.6}$$

The $(2\pi)^6$ factor has been omitted; the sum of the squares of the $2l + 1$ spherical harmonics gives 1.

Since the deuteron wave function $d(\mathbf{r}_p - \mathbf{r}_n) = d(|\mathbf{r}_p - \mathbf{r}_n|)$ is rather flat, its Fourier transform D drops rather rapidly with increasing

$$|\mathbf{k}_e - \tfrac{1}{2}\mathbf{k}_i| = (k_e^2 + \tfrac{1}{4}k_i^2 - k_e k_i \cos \theta)^{1/2}, \tag{10.7a}$$

where θ is the angle between incident deuterons and emitted protons. As a result, most of the protons are emitted at low θ in the forward direction. The last factor of (10.6) tends to decrease this effect.

The kinetic energy of the neutrons in the bound orbit u is, in general, much higher than in the deuteron. As a result, U_r drops much less rapidly with increasing argument than D. On the other hand, the wave function in momentum space $U(\mathbf{k})$ has a similar behavior to that of the coordinate wave function: it has an l-fold zero at $k = 0$. Hence, unless the neutron is captured into an s orbit ($l = 0$), the U_r will be very small for small

$$|\mathbf{k}_e - \mathbf{k}_i| = (k_e^2 + k_i^2 - 2k_e k_i \cos \theta)^{1/2}, \qquad (10.7\mathrm{b})$$

i.e. for small θ. This depression at low θ is the more pronounced the larger is l. Naturally, for very large l the whole cross-section (10.6) remains very small: for low θ the first, for high θ the second factor becomes very small. For intermediate l, however, the maximum of $J(\mathbf{k}_e)$ wanders to higher and higher θ, decreasing at the same time in absolute magnitude. The angular distribution calculated by Butler for various θ, taking the spins of the deuteron and the proton and neutron into account, are illustrated in Fig. 10.2. Comparison of the observed angular distributions with these curves permits a determination of the angular momentum of the orbit into which the neutron is captured. Similarly, the angular distribution of the neutrons from the (d, n) reaction permits the determination of the l value of the captured proton.

The preceding calculation reproduces the characteristic features of the observed angular distribution. In order to obtain quantitative agreement, it was found necessary to restrict the integration over $\mathbf{r}_n$ in (10.3) to the outside of the nucleus so that only the tail of $u(\mathbf{r}_n)$, which extends beyond the nuclear radius as usually considered, plays a role. This shows that the reaction takes place on the surface of the nucleus, and can be interpreted most naturally by assuming that the interaction between the nucleus and the incident deuteron distorts the wave function of the latter in such a way that its neutron does not enter the nucleus. This is reasonable because the stripping distribution is obtained only at deuteron energies *between* resonance levels, and the incident deuteron does not penetrate the nucleus at these energies. This appears natural on the basis of the qualitative picture given in Section 9.2 and follows also from the more quantitative theory of 9.3. The deuterons, the energy of which corresponds to a resonance level of the compound nucleus, on the other hand, will penetrate the nucleus, but the angular distribution of the resulting protons and neutrons will be characterized by the angular momentum of the compound state. It would consist of a single $P_l(\theta)^2$ were it not for the spins of nucleus, deuteron, and proton. On account of these it becomes the square of a linear combination of a few P_l. At resonance the reaction ceases to be a surface reaction and is not expected to show the stripping distribution. In the neighborhood of the resonance the interference of stripping and resonance amplitudes should give rise to interesting phenomena which have not been investigated fully.

The $V_p(r_p)$ term of (10.3) needs yet to be considered. The contribution of the $V_p(r_p)$ term corresponds to processes in which the proton is pushed away from the neutron by the nuclear potential and the neutron

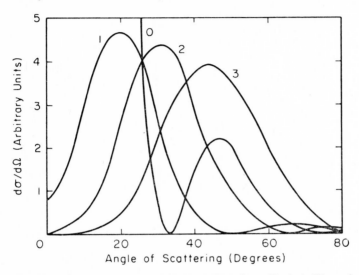

Figure 10.2. Angular Distribution of Stripping Reactions (Butler). The calculated angular distribution of the protons from a (d, p) reaction is given for various l values of the captured neutron.

finds itself captured. The probability of such a process is considerably smaller than the one calculated above. The term with $V_p(r_p)$ can indeed be calculated to give:

$$(2\pi)^{3/2} \int v_p(\mathbf{k}_i + \mathbf{k} - \mathbf{k}_e) U(\mathbf{k}) D(\mathbf{k} + \tfrac{1}{2}\mathbf{k}_i) \, d\mathbf{k}, \tag{10.8}$$

in which v_p is the Fourier transform of V_p. Because of the behavior of D as outlined above, the integrand is large only for $\mathbf{k} + \tfrac{1}{2}\mathbf{k}_i \approx 0$. Hence, one can write for (10.8) approximately:

$$(2\pi)^{3/2} v_p(|\tfrac{1}{2}\mathbf{k}_i - \mathbf{k}_e|) U(\tfrac{1}{2}\mathbf{k}_i) \int D(\mathbf{k} + \tfrac{1}{2}\mathbf{k}_i) \, d\mathbf{k}$$
$$= (2\pi)^3 v_p(|\tfrac{1}{2}\mathbf{k}_i - \mathbf{k}_e|) U(\tfrac{1}{2}\mathbf{k}_i) \, d(0). \tag{10.9}$$

This term does not show the characteristic behavior of (10.6). However, the $d(0)$ factor renders it considerably smaller than the latter, and there is reason to believe that higher approximations further reduce its magnitude. Hence, (10.6) appears a reasonable expression for the angular distribution: the possibility to neglect the $V_p(r_p)$ term of (10.3) is hardly surprising since the neutron capture must be due, after all, principally to the interaction of the neutron, rather than that of the proton, with the nucleus.

Butler's method of determining the orbit into which a particle is captured often leads to a determination of the angular momentum and parity of the product nucleus. On this account, it is a particularly significant example to illustrate an approximate method for the description of nuclear collisions of a rather general kind.

10.2. Electric Excitation

In the collision of a charged particle with a nucleus, inelastic scattering and consequently nuclear excitation may occur even if specific nuclear forces do not come into play. The phenomenon may be described as follows. In the motion of the incident charged particle past the target nucleus, the latter is subjected to a rapidly varying electric field. This field may give rise to transitions from the ground state to excited states of the target nucleus. The process is thus a species of photoeffect, and is referred to as electric or Coulomb excitation. A theoretical description of the process was first given by Weisskopf (1938), but only in recent years has it been definitely observed. The cross-section for electric excitation (about 10^{-28} or 10^{-29} cm^2) is small compared with direct nuclear transformations, and is observable only when the latter are extremely improbable. This is the case, for example, in the collisions of protons or α particles of intermediate energy on heavy nuclei. In such cases the electrostatic repulsion between the colliding particles is so large that the classical distance of closest approach is large compared with nuclear radius; the probability of direct nuclear interaction is, therefore, extremely small. When the direct nuclear transformation may be disregarded, the electric excitation is detected by observing the γ-radiation emitted in the de-excitation of the target nucleus.

To calculate the cross-section for electric excitation, we may neglect the nuclear forces between incident and target particles and write for the Hamiltonian of the system:

$$H = T_r + H_A + Ze^2 \sum_p \frac{1}{|\mathbf{r} - \mathbf{r}_p|}, \qquad (10.10)$$

where T_r is the kinetic energy of relative motion; H_A, the Hamiltonian of the target nucleus; Z, the charge number of the incident particle; $\mathbf{r}$, the vector from the center of mass of the target nucleus to the incident particle, and the $\mathbf{r}_p$, the positional coordinates of the protons of the target. The electrostatic energy term is

$$Ze^2 \sum \frac{1}{|\mathbf{r} - \mathbf{r}_p|} = \frac{ZZ'e^2}{r} + Ze \sum_{l=1}^{\infty} \sum_m \frac{4\pi}{2l+1} Q_{lm} r^{-l-1} P_{lm}(\theta, \phi), \quad (10.11)$$

where Z' is the charge number of the target nucleus, θ, ϕ and θ_p, ϕ_p are the polar angles of $\mathbf{r}$ and $\mathbf{r}_p$, respectively, and:

$$Q_{lm} = e \sum_p r_p^l P_{lm}^*(\theta_p, \phi_p). \quad (10.11a)$$

The operators Q_{lm} are known as the electric multipole operators; l defines the multipole order of the operator (electric dipole for $l = 1$, electric quadrupole for $l = 2$, etc.). These operators occur also in the description of spontaneous electromagnetic transitions. The Hamiltonian H will lead to simple elastic Rutherford scattering if only the r^{-1} term in the potential energy is retained. The second term in the potential energy designated below as V' may be considered as providing transitions between simple Rutherford scattering states with excitation of the target nucleus.

The probability of excitation may be obtained by means of first order perturbation theory. The specifically nuclear contribution to the excitation probability occurs in the form of a matrix element of V' between the initial and final target states. In general, only the multipole operators of one particular order will contribute strongly to the matrix element (usually the lowest l value for which the matrix element of Q_{lm} is not zero), so that we may speak of electric dipole excitations (E-1), electric quadrupole excitations (E-2), etc. In the case of electric l-pole excitation, the only *nuclear* factor in the cross-section is:

$$B_l = \sum_{m M_i M_f} |(\psi_f, Q_{lm} \psi_i)|^2, \quad (10.12)$$

where ψ_i and ψ_f are the initial and final nuclear states and M_i and M_f are the possible magnetic quantum numbers for these states. This same factor occurs in the rate of radiative transitions of polarity l.

The absolute value of the cross-section has been calculated accurately for dipole transitions by considering the second term of the multipole expansion (10.11) to be a small perturbation. This is surely justified. An approximate expression for the motion of the incident particle under the influence of the Coulomb term—that is, the first term of (10.11)—has been used for the calculation of the E-2 and higher multipole transitions. It has been shown, however, that the approximation involved is very good.

Experiments on the angular distribution of the γ-radiation emitted in the process lead to information on the spin values of the nuclear states involved. Measurement of the excitation cross-section as a function of incident energy is useful for the determination of the multipole order responsible for the process. A further measurement of the absolute probability of excitation makes possible the evaluation of B_l.

The different nuclear models imply rather different values for the quantities B_l. The single particle and collective models contrast rather strongly in this respect. The collective model provides relatively large values for B_2 (compared with those from a pure single particle model) for transitions between the ground states and the low-lying "rotational levels" of heavy nuclei. The estimates of B_2 obtained by electric excitation lend support to the collective model.

CHAPTER 11

Interaction with Electron-Neutrino Fields

11.1. Theory of β-Decay

A qualitative description of the phenomenon of β-decay is given in Chapter 1. Here a somewhat more detailed account of the theory and of its experimental verification is given. Fermi's theory of the β-process is patterned on the theory of radiation. In the radiative process we envisage a change of state of the charged radiating system accompanied by the emission of a photon; the β-decay is pictured as a change of the state of a nucleon (or of a nuclear system) together with the emission of an electron and a neutrino. In the case of radiation, the form of the interaction between the charged particles and field is taken over from classical theory; the interaction strength is measured by the electronic charge. For the β-decay an interaction form must be invented which is consistent with experimental observations.

Fermi's theory is formulated in terms of the operators of quantum field theories. The β-decay consists either of the annihilation of a neutron and of a neutrino and the creation, instead, of a proton and of an electron, or of the inverse process. It is described, therefore, by two products of four field operators of the form:

$$\psi_p^\dagger \psi_n \psi_e^\dagger \psi_\nu + \psi_\nu^\dagger \psi_e \psi_n^\dagger \psi_p \tag{11.1}$$

The indices p, n, e, ν refer to protons, neutrons, electrons, and neutrinos; the ψ are annihilation; the $\psi^\dagger$, creation operators. The latter are the hermitean conjugate of the former.

The particles can be created and annihilated at any point of space and with any of the four values of Dirac's spin variable: all the $\psi = \psi(\mathbf{r}, \zeta)$ depend on position and on the value of the spin variable. The total interaction is an integral of expressions of the form (11.1) over all space, and could be an arbitrary linear combination of the 256 components which can be obtained by giving all four spin variables every possible value:

$$H_{\text{interaction}} = \sum_{\zeta_p \zeta_n} \sum_{\zeta_e \zeta_\nu} C(\zeta_p, \zeta_n, \zeta_e, \zeta_\nu) \int \psi_p^\dagger(\mathbf{r}, \zeta_p) \psi_n(\mathbf{r}, \zeta_n) \tag{11.1a}$$

$$\cdot \psi_e^\dagger(\mathbf{r}, \zeta_e) \psi_\nu(\mathbf{r}, \zeta_\nu) \, d\mathbf{r}$$

$$+ \text{ hermitean-adjoint.}$$

The "hermitean adjoint" corresponds to the second term of (11.1).

Certainly (11.1a) is the simplest expression which accounts for the possibility of the β-transformation at any point of space.

A condition on the coefficients C follows from the requirement of relativistic invariance of the interaction. There are sixteen products of the form $\psi_e^\dagger(\mathbf{r}, \zeta_e)\psi_\nu(\mathbf{r}, \zeta_\nu)$ corresponding to the four values of ζ_e and ζ_ν each. From these sixteen products one can form the following linear combinations: one scalar (S), the four components of a vector (V), the six components of an antisymmetric tensor (T), four components of an axial vector (A), and a pseudoscalar (P). The same linear combinations can be formed from the sixteen products $\psi_p^\dagger(\mathbf{r}, \zeta_p)\psi_n(\mathbf{r}, \zeta_n)$. The contraction of the quantities formed from the products $\psi_e^\dagger\psi_\nu$ with the corresponding quantities formed from the $\psi_p^\dagger\psi_n$ gives five invariants (scalars). The corresponding operators are designated by H_K (with $K = S, V, T, A, P$). Thus, the most general relativistically invariant interaction of the form (11.1a) can be written as:

$$H_{\text{interaction}} = \sum g_K H_K, \qquad (11.2)$$

where the g_K (which are real) characterize the strengths of the corresponding interactions H_K. Comparison of the theoretical consequences of the hypothesis (11.2) with experimental observation tests the adequacy of the hypothesis, and if adequate, permits the measurement of the various strengths g_K. Fermi's original assumption made use only of H_S. Gamow and Teller later suggested that the interaction was of the type H_T or H_A. It is now known that the interaction must be a combination of both Fermi and Gamow-Teller types. Data on β-decay are well accounted for by the interaction (11.2). They are consistent with the form:

$$H_{\text{interaction}} = g_S H_S + g_T H_T, \qquad (11.2\text{a})$$

or with the form:

$$H_{\text{interaction}} = g_V H_V + g_A H_A, \qquad (11.2\text{b})$$

with $g_S = 1.40 \times 10^{-49}$, $g_T = 2 \times 10^{-49}$ erg cm^3; in the latter case these numbers correspond to g_V and g_A, respectively. The coefficients g_V and g_A amount to at most a few percent of the coefficients g_S and g_T if (11.2a) is valid. Conversely, if (11.2b) is valid, g_S and g_T are very small. Because of our inability to compute the relevant nuclear matrix elements, we do not have a measure of the strength of the pseudoscalar (P) type of interaction. The probability of the transition from the nuclear state Ψ_i to the nuclear state Ψ_f, which is accompanied by the emission of an electron-antineutrino or positron-neutrino pair with momenta $\mathbf{p}$ and $\mathbf{q}$, can be calculated by means of the general theory of transition probabilities. Since the original and final states of the neutrino and electron fields are well known—they correspond to states of the free particle or to particles under the influence of a Coulomb field—the integration over the coordinates of these particles can be carried out. The resulting expression for the transition probability contains a matrix element which refers only to the nuclear states Ψ_f and Ψ_i. It is:

$$\lambda(f, i; \mathbf{p}, \mathbf{q}) = \frac{2\pi}{\hbar} \sum_{Km} |(\Psi_f, g_K Q_{Km} \Psi_i)|^2 \rho \qquad (11.3)$$

where ρ is the density of the states of the electron-neutrino field. If one neglects the effect of the Coulomb field on the electron, the operators Q_{Km} become

$$Q_{Km} = \sum_{j=1}^{A} q_{Km}(j) \tau_{j\xi} \exp -i(\mathbf{p} + \mathbf{q}) \cdot \mathbf{r}_j. \qquad (11.3a)$$

The sum over j appears because each of the nucleons may, potentially, undergo a transformation. The exponential corresponds to the recoil of the transforming particle which is oppositely equal to the momentum of the electron-neutrino pair. The operator $\tau_{j\xi}$ is the first component of the isobaric spin operator of particle j; it transforms a proton into a neutron and conversely. The operators $q_{Km}(j)$ apply only to the spin coordinate ζ_j of particle j; they are essentially the matrices which form the component m of the quantity K from the products $\psi_p^\dagger(\mathbf{r}, \zeta_p) \psi_n(\mathbf{r}, \zeta_n)$. They are α-matrices of Dirac or products of such matrices; their form is not essential for what follows.

The expressions (11.3) and (11.3a) imply that the nuclear wave functions Ψ_f and Ψ_i depend on four-valued spin variables rather than the two-valued spin variables used hitherto. Hence, in order to evaluate these expressions, it would be necessary to convert the ordinary nuclear wave functions with two-valued spin variables to nuclear wave functions with four-valued spin variables. This is possible only with a relativistic theory of nuclear interaction. Since such a theory is lacking, calculation of the matrix elements in (11.3) has a certain amount of ambiguity, even if Ψ_i and Ψ_f are known as functions of the positional and two-valued spin coordinates. This is the principal reason for the difficulty of determining the strength g_p of pseudoscalar interaction. For the other types of interaction (11.3) can be evaluated at least approximately because the velocity of the nucleons is, on the whole, small compared with that of light. As a result, the wave functions Ψ_i and Ψ_f are large only for two values of each spin variable ζ_j, and these two values can be identified with the two values of the ordinary spin variable.

Some of the matrices q_{Km} connect large components of Ψ_i with large components of Ψ_f (and small components with small components). The order of magnitude of these matrix elements is 1; they will be denoted by $q_{Km}^{(n)}$ (n for non-relativistic). They can be obtained with a reasonable accuracy if the large components of Ψ_i and Ψ_f are known, that is, if these are known as a function of the two-valued spin variables. Other q_{Km} connect large components of Ψ_i with small components of Ψ_f, and vice versa. The order of magnitude of these is v/c, where v is the average velocity of the nucleons in the nucleus; they are denoted by $q_{Km}^{(r)}$ (r for relativistic). The pseudoscalar interaction's matrix $(\alpha_x\alpha_y\alpha_z\alpha_t)$ is in the second category, and these matrix elements could be determined only if the small components of Ψ_i and Ψ_f were known, which is not the case.

11.2. Allowed and Forbidden Transitions

In order to obtain an estimate for the matrix element in (11.3) it is useful to expand the exponential:

$$\exp -i(\mathbf{p}+\mathbf{q})\cdot\mathbf{r}/\hbar = \sum i^{-l} \mathcal{E}^{(l)}; \qquad \mathcal{E}^{(l)} = \frac{1}{l!}\left(\frac{(\mathbf{p}+\mathbf{q})\cdot\mathbf{r}}{\hbar}\right)^l. \quad (11.4)$$

Since the order of magnitude of p and q is mc, with m the electronic mass, and since only such r contribute significantly to the integral (11.3) which are not larger than the nuclear radius, every term $\mathcal{E}^{(l)}$ of (11.4) gives, in general, a smaller contribution than the preceding one by a factor of the order of magnitude 10^{-2}—unless the integral of the preceding $\mathcal{E}^{(l)}$ vanishes on account of a selection rule or some similar reason. It is customary, therefore, to replace the operators Q_{Km} by the operators of lowest l,

$$M_{Km}^{(l)} = \sum_j [q_{Km}^{(n)}(j)\tau_{j\xi}\mathcal{E}^{(l)} + q_{Km}^{(r)}(j)\tau_{j\xi}\mathcal{E}^{(l-1)}], \quad (11.4a)$$

for which the integral (11.3) does not vanish. The two terms of (11.4a) are of similar order of magnitude and their selection rules also have much in common, so that if the matrix element of the second operator does not vanish, that of the first one is also finite. In particular, the $q^{(n)}$ do not change the parity of a state, while $\mathcal{E}^{(l)}$ changes it for odd l, leaves it unchanged for even l. Hence the first term may give a finite contribution only if the parities of Ψ_i and Ψ_f are the same and l is even, or if the parities of Ψ_i and Ψ_f are opposite and l is odd. This is true for all $\mathbf{p}$ and $\mathbf{q}$. The same is true of the second term of (11.4a) also because the $q_{Km}^{(r)}$ are those Dirac matrices which do change the parity. For $l = 0$ there is no second term in (11.4a). As the form of $\mathcal{E}^{(l)}$ indicates, successively higher l permit an increasing difference in the total angular momenta J of Ψ_i and Ψ_f. If the $l = 0$ term can give a finite contribution to the matrix element, the transition is called allowed; if $M_{Km}^{(l)}$ is the first term which gives a finite contribution, the transition is called l'th forbidden. Table 11.1 gives the selection rules for all types K of interaction in the case of allowed transitions and the other selection rules for S and T transitions. Note that P gives no allowed transition.

TABLE 11.1

Order of Forbiddenness	Type Interaction (K)	Change in J	Change in Parity
allowed	SV TA	0 0 or 1 but not $0 \to 0$	no
1st forbidden	S T P	0,1 0,1,2 0	yes
2nd forbidden	S T	2 2,3	no
3d forbidden	S T	3 3,4	yes

For the forbidden transitions, only those changes of J are given which did not occur for the same interaction at a lower degree of forbiddenness. We note further that $M_S^{(0)}$ and $M_V^{(0)}$ are, essentially:

$$M_S^{(0)} \approx M_V^{(0)} \approx \sum \tau_{i\xi} = 2T_\xi, \qquad (11.5)$$

and that $M_T^{(0)} \approx M_A^{(0)}$ has three components:

$$\sum \sigma_{ix}\tau_{i\xi} = 2Y_{x\xi}; \qquad \sum \sigma_{iy}\tau_{i\xi} = 2Y_{y\xi}; \qquad \sum \sigma_{iz}\tau_{i\xi} = 2Y_{z\xi}. \quad (11.5\text{a})$$

11.3. Shape of the Spectrum

The expressions (11.5), (11.5a) for the matrix elements of allowed transitions do not depend on the momentum of the emitted electron and neutrino. As a result, the shape of the allowed spectra is entirely determined by the factor ρ in (11.3). This depends, in general, on the momentum and polarization of electron and neutrino. To obtain the energy spectrum of the electrons one has to sum over all directions of polarization and all directions of momenta. The expression obtained is proportional to the volume in phase space per unit energy range available for electron and neutrino. This is

$$P(W_e) = pW_e qW_q, \qquad (11.6)$$

where p and q are, as before, the momenta of electron and neutrino; W_e and W_q, the corresponding energies, including rest-mass. The sum of these is the energy of disintegration $W_0 = W_e + W_q$, so that (11.6) contains only one independent variable. It gives the spectral distribution of the electrons, that is, apart from a constant factor, the number of electrons the energy of which is W_e within unit energy interval.

If the rest-mass of the neutrino is zero, $W_q = cq$ and (11.6) becomes:
$$P(W_e) = pW_e(W_0 - W_e)^2/c. \tag{11.6a}$$

In this case the upper limit of the electron energy is W_0 and the spectral distribution (11.6a) behaves like $(W_0 - W_e)^2$ in the neighborhood of the upper limit. If the mass m_ν of the neutrino is finite, the upper end of the spectrum is at $W_{\max} = W_0 - m_\nu c^2$, and the spectral distribution (11.6) is proportional to $(W_{\max} - W_e)^{1/2}$ in the neighborhood of the end-point. Hence the electron spectrum depends, in the neighborhood of its end-point, on the neutrino mass in a sensitive way. The lowest limit for the rest-mass of the neutrino was obtained by careful investigation of the electron spectrum near the end-point.

The effect of the electrostatic field of the nucleus on the wave function of the electron is not taken into account in the preceding calculation. To take it into account, the plane wave $\exp(i\mathbf{p}\cdot\mathbf{r})$ in (11.3a) should be replaced by a Coulomb wave function. This introduces a correction factor into the spectral distribution (10.6a) which is usually denoted by $F(Z_-, W_e)$ for electrons and by $F(Z_+, W_e)$ for positrons. These correction factors are not very important for allowed transitions and low Z unless the energy of disintegration is very small. They do not influence the behavior of the spectral distribution in the neighborhood of the end-point.

The spectral distribution (11.6a) of the electrons with the correction factor $F(Z_\pm, W_e)$ for the Coulomb field was confirmed by very careful and accurate measurements. These form a strong support of Fermi's theory. The spectral distribution of the electrons or positrons contains a polynomial of l'th or $(l-1)$'th degree of p and q for transitions forbidden to the l'th degree. Such factors appear in the expression (11.4) for $\mathcal{E}^l$. In some cases therefore the degree of forbiddenness can be obtained from the electron spectrum directly. This is particularly true for certain second-forbidden transitions. The influence of the Coulomb field gives most first-forbidden transitions the allowed shape. The degree of forbiddenness manifests itself, as a rule, also in the absolute rate of the transition.

11.4. Total Transition Probability

The total transition probability, that is, the reciprocal mean lifetime of a β-active nucleus, is obtained by integrating (11.3) over all possible momenta $\mathbf{p}$ and $\mathbf{q}$ of electron and neutrino. For allowed transitions the matrix element is practically independent of $\mathbf{p}$ and $\mathbf{q}$, and the integration over the directions of $\mathbf{p}$ and $\mathbf{q}$ was indicated before. Hence, the total transition probability is the square of the matrix element which occurs in (11.3), multiplied with the integral over the spectral function $P(W_e)$, together with the correction factor $F(Z_\pm, W_e)$. This integral has to be extended over the whole range of the electron spectrum—that is, from mc^2 to W_0. The result of the integration depends, in addition to the energy of disintegration W_0, also on the charge number Z of the product nucleus, though in a less sensitive way. The result of the integration is usually denoted by $f(Z_*, W_0)$. Hence, the disintegration constant for allowed transitions becomes

$$\lambda = \frac{f(Z_*, W_0)}{2\pi^3 \hbar^7 m^{-5} c^{-4}} \{g_S^2 |T_\xi|^2 + g_T^2 |\mathbf{Y}_\xi|^2\} = \frac{\ln 2}{t}. \quad (11.7)$$

The denominator comes from factors which we did not carry along; t is the half-life of the nucleus. In (11.7) it is assumed that only scalar and tensor interactions are present. The indicated matrix elements are those for the operators defined in (11.5) and (11.5a). $|T_\xi|^2$ and $|\mathbf{Y}_\xi|^2$ are abbreviations for:

$$|T_\xi|^2 = |(\Psi_f, 2T_\xi \Psi_i)|^2 \quad \text{and} \quad (11.7a)$$

$$|\mathbf{Y}_\xi|^2 = |(\Psi_f, 2Y_{x\xi} \Psi_i)|^2 + |(\Psi_f, 2Y_{y\xi} \Psi_i)|^2 + |(\Psi_f, 2Y_{z\xi} \Psi_i)|^2. \quad (11.7b)$$

The product ft depends, apart from well known universal constants, only on the strengths of the interaction and the nuclear matrix elements of T_ξ and $Y_{x\xi}$, $Y_{y\xi}$, $Y_{z\xi}$;

$$ft = \frac{\text{Const.}}{g_S^2 |T_\xi|^2 + g_T^2 |\mathbf{Y}_\xi|^2}. \quad (11.8)$$

The experimental ft values fall into two groups: the "favored" or "superallowed" group with log ft ranging from 2.9 to 3.7, and an "unfavored" or "normal" group with log ft mainly in the range from 5 to 6. In some exceptional cases log ft for allowed transitions fall outside the range above; e.g. for the decay of C^{14}, which is an allowed transition, log $ft = 9$. The favored group consists of the transitions between mirror nuclei and a group of transitions between even-even $|T_{\bar\xi}| = 1$, and odd-odd $T_{\bar\xi} = 0$ nuclei, such as $He^6 \rightarrow Li^6$, $C^{10} \rightarrow B^{10}$, etc.

In principle, g_S^2 and g_T^2 could be obtained by means of (11.8) from the measured half-lives and the $|T_\xi|$ and $|\mathbf{Y}_\xi|$ matrix elements for two decays. It is useful to consider the favored decays, since for these cases the matrix elements must be close to their maximum possible values. For the light nuclei which alone exhibit the favored transitions, the isobaric spin quantum number T is probably good, and on this assumption the T_ξ matrix element can be computed; it vanishes except between states of the same isobaric spin multiplet. The calculation of $|\mathbf{Y}_\xi|^2$, however, requires a more detailed knowledge of the nuclear states involved. This matrix element can be computed on the basis of particular nuclear models and, as was mentioned before, comparison with experimental data favors, on the whole, the supermultiplet model, at least for light nuclei and the early parts of the shells. However, the values obtained for g_T^2 from different transitions are not too closely equal in any model, reflecting, presumably, the inadequacy of these models for the accurate calculation of $|\mathbf{Y}_\xi|$. The statistical analysis of all transitions for which the model employed for the calculation of $|T_\xi|$ is considered adequate gives the values for these quantities mentioned in Section 11.1.

CHAPTER 12

Electromagnetic Transitions in Complex Nuclei

12.1. Introduction

Until relatively few years ago the investigation of electromagnetic transitions between nuclear states played a small role in nuclear physics. The reasons for this were mainly technical. The recent developments of the scintillation counter, fast electronics, and the methods of investigating resonance scattering, together with theoretical calculations on internal conversion coefficients, angular correlation, and Coulomb excitation (Chapter 10) have made it possible to derive useful data for the understanding of nuclear structure from the study of radiative transitions. The data provide inferences of effectively two types. Measurement of internal conversion coefficients and of angular distributions and correlations lead to inferences on spin values and parities of excited states of nuclei. Measurements of lifetimes or widths of absorption lines, on the other hand, give values of the absolute squares of matrix elements of transition operators. If the interaction responsible for the electromagnetic transition arises wholly from the free charges and dipole moments of the nucleons, the measured matrix elements in question are off-diagonal elements of operators similar in structure to the magnetic dipole and electric quadrupole operators which were considered in connection with static moments of the ground states of nuclei (Chapter 3) and more explicitly in connection with Coulomb excitation (Chapter 10) and β-decay (Chapter 11).

Insofar as problems of nuclear structure are concerned, the study of radiative transitions and the study of β-decay play similar roles. The relative ease of detection of the β-particles and of the measurement of the slow transition rates stimulated the early study of the β-decay processes. Analysis of the rate of β-transitions has theoretical advantages also because the operators responsible for these are the most simple operators in the case of allowed transitions. Hence inferences from the measured rates to the properties of nuclear states are fairly immediate. Analysis of nuclear states by means of β-decay suffers, however, from two features: (a) the slow transition rate for the process means that for the most part only the ground states of the parent nuclei are involved, and (b) in the case of forbidden transitions, which form the majority of all transitions, matrix elements of several operators play a role. The radiative transitions connect states of a single nucleus. The operator responsible for the simplest transitions involves more complex properties of nuclear states than do the allowed β-decay operators. On the other hand, the radiative transition rates are probably easier to interpret than the rates of forbidden β-transitions. This applies at least to the more important class of transitions, the electric transitions. The situation may not be so simple as far as magnetic transitions are concerned.

An excited nucleus may lose its energy either by simple radiation or by internal conversion. In the latter process the excitation energy is transferred to an extra-nuclear electron. Internal conversion is of particular importance if the energy of the transition is low (< 1 Mev), and the charge of the nucleus is high, or if the radiative transition is inhibited by special selection rules (e.g. in 0 → 0 transitions, i.e. transitions between states of zero angular momentum). In addition, if the energy of the transition is large enough, de-excitation may occur through production of pairs. In general this process is considerably slower (about 10^{-3} times slower) than the radiative process and has little influence on the effective half-life of the excited state. However, the 6.06 mev 0 → 0 transition in O^{16} occurs almost wholly by pair production.

In the following the theories of radiative transition and of internal conversion are sketched in a qualitative manner. The measured matrix elements are then compared in a general way with theoretical estimates based on the single particle model.

12.2. Radiative Transitions

The theory of electromagnetic radiation is much older than that of β-decay and the literature contains good comprehensive treatments of the subject. In fact, as mentioned before, the theory of β-decay was largely patterned on the quantum theory of radiation. The following treatment, which is now the customary one, differs from that adopted for β-decay inasmuch as it uses spherical rather than plane waves as the basic set of states for the electromagnetic field. Such spherical waves have the advantage that they possess, in addition to a definite energy, also a definite total angular momentum, which will be called l, and even or odd parity. The total angular momentum is usually specified as the multipolarity of the radiation. Instead of a direct specification of parity the "electric" (E) or "magnetic" (M) character of the radiation is given. The parity of a radiation state is then defined by l and the character of the radiation (E or M). Electric radiation has parity $(-)^l$, while magnetic radiation has parity $(-)^{l+1}$. The radiation field has, for a given l and parity, $2l + 1$ different states characterized by $2l + 1$ different values ($\mu = -l, -l + 1, \cdots, l - 1, l$) of the z component of total angular momentum. The names "electric" and "magnetic" derive from the classical distribution of charges and currents which give rise to these fields. Expansion of the field into spherical rather than plane waves is preferable if only one particle (in this case a γ-quantum) is emitted,

12.2 · ELECTRIC AND MAGNETIC MULTIPOLES

because energy, angular momentum, its z component μ, and parity completely specify the state. This is not the case if two particles are emitted. Use of spherical waves, in the case of the emission of two particles, would obscure the angular correlations between these.

The rate of radiative transitions is calculated, as that of β-decay, by the theory of transition probabilities. Since the state of the emitted photon is completely determined by its energy and character, the probability of emission is given by an expression analogous to that of (11.3) for β-decay:

$$\lambda(f, i; l, \mu, E) = \frac{2\pi}{\hbar} |(\Psi_f, eQ(l, \mu, E)\Psi_i)|^2 \rho. \qquad (12.1)$$

Since the strength and form of the operator of electromagnetic interaction are well known, the indeterminate constant g_K of β-decay could be replaced by the charge of the electron and there is no need for the distinguishing marks K, m used in the case of β-decay. Because of the different choice of basic states, the momenta p and q of electron and neutrino are replaced by the characteristics of the emitted photon. The E in (12.1) indicates an electric transition but there is a similar expression, with M instead of the E, for magnetic transitions. The energy of the photon is given as the energy difference between initial and final states; ρ is again the number of states of the photon per unit energy interval.

The wave lengths of the γ-quanta are much longer than the radius of the nucleus. One can, therefore, expand the operator Q in (12.1) as function of the radius and discard all but the first term. This gives:

$$\lambda(f, i; l, \mu, E) = \frac{8\pi(l-1)!(l+1)!2^l}{[(2l+1)!]^2} k^{2l+1} |E_{l\mu} + E'_{l\mu}|^2, \qquad (12.2)$$

and a similar expression for magnetic transitions in which E is replaced by M. The wave number k of the photon is its energy divided by $\hbar c$. The matrix elements E and M are:

$$E_{l\mu} = e \sum_{j=1}^{z} (\Psi_f, r_j^l Y_{l\mu}(\theta_j, \phi_j)^* \Psi_i) \quad \text{and} \qquad (12.3a)$$

$$M_{l\mu} = \frac{e\hbar}{Mc} \frac{1}{l+1} \sum_{j=1}^{z} \sum_{\alpha} \left(\Psi_f, \frac{\partial}{\partial x_{j\alpha}} [r_j^l Y_{l\mu}(\theta_j, \phi_j)]^* \mathbf{L}_{j\alpha} \Psi_i\right). \qquad (12.3b)$$

These expressions represent the interaction of the charge distribution and of the current due to the motion of the protons with the electromagnetic field. Hence, the summation is extended only over the coordinates of the protons. Alternately, a factor $\frac{1}{2}(1 - \tau_{jt})$ could be introduced and the summation extended over all particles. M is the mass of the proton; r_j, θ_j, ϕ_j, its polar, and x_j, its rectangular coordinates: the summation over α is to be extended over the three coordinates x, y, z. The $\mathbf{L}_j$ is the operator of the angular momentum:

$$\mathbf{L}_j = -i r \times \operatorname{grad}_j ; \tag{12.4}$$

the Y are normalized spherical harmonics. The interaction of the spin with the electromagnetic field is represented by the E' and M' terms. Only

$$M'_{l\mu} = \sum_{j=1}^{A} \sum_{\alpha} \mu_j \left(\Psi_f, \frac{\partial}{\partial x_{j\alpha}} [r_j^l Y_{l\mu}(\theta_j, \phi_j)]^* \sigma_{j\alpha} \Psi_i \right) \tag{12.3c}$$

is important; $E'_{l\mu}$ is in general small as compared with $E_{l\mu}$. The summation here extends over all particles; μ_j is the magnetic moment of particle j (that is 2.78 $e\hbar/2Mc$ for protons and -1.91 $e\hbar/2Mc$ for neutrons). For $l = 1$, $\mu = 0$, that is the z component of the dipole radiation, one obtains the expressions:

$$E_{10} = e \left(\frac{3}{4\pi} \right)^{1/2} \sum_{j=1}^{Z} (\Psi_f, z_j \Psi_i), \tag{12.4a}$$

$$M_{10} = \frac{ie\hbar}{2Mc} \left(\frac{3}{4\pi} \right)^{1/2} \sum_{j=1}^{Z} \left(\Psi_f, \left(y_j \frac{\partial}{\partial x_j} - x_j \frac{\partial}{\partial y_j} \right) \Psi_i \right), \tag{12.4b}$$

$$M'_{10} = \left(\frac{3}{4\pi} \right)^{1/2} \sum_{j=1}^{A} \mu_j (\Psi_f, \sigma_{jz} \Psi_i). \tag{12.4c}$$

The selection rules can be read off these expressions immediately, but they are independent of the assumption that the wave length of the emitted light is much longer than the radius of the nucleus. If the parities of Ψ_i and of Ψ_f are the same, only even electric multipole and odd magnetic multipole radiations are possible. If the parities of Ψ_i and of Ψ_f are opposite, the polarity of the electric radiation is odd; that of the magnetic radiation, even. In addition, the J values of Ψ_i and Ψ_f must form a vector triangle: $|J_i - J_f| \leq l \leq J_i + J_f$. Transition rates of the electric multipole radiations decrease with increasing l and the same is true of transition rates due to magnetic multipole radiations. Hence, if the selection rules admit an electric l-pole radiation, transition rates due to higher electric multipoles can be, in general, disregarded, and the same is true of the magnetic multipoles. In general, the magnetic l-pole should give a lower rate than the electric l-pole; in the case of dipole radiation the two are often of the same order of magnitude. It is more usual, however, to consider magnetic l-pole and electric $(l + 1)$-pole to be of the same order of magnitude; they also can occur in the same transition and often compete with each other. If J_i and J_f differ by several units, the multipole radiation will be of high order and the transition slow. This is the explanation of nuclear isomerism, suggested by v. Weizsäcker (Chapter 1).

Expressions for radiative transition rates are used, principally, to test the accuracy of wave functions furnished by the various models. They are given already in terms of the wave functions with two-valued spin; this accounts for the fact that they are rather artificially divided into two parts: one due to charges, the other due to spin. There is general agreement that the expressions for the electric transition rates are sufficiently accurate. The rate of the magnetic transitions, on the other hand, may be influenced by the currents which are not due to motion of the nucleons but to the mesons which transmit the nuclear interaction. These currents are held responsible also for the magnetic moment of the neutron and the anomalous magnitude of the magnetic moment of the proton. The use of the observed moments μ_i in (12.3c) therefore already partially accounts for these "exchange currents". It is to be expected, however, that the meson current associated with the free nucleons is modified by the proximity of other nucleons so that appreciable corrections to the transition rates may result. The magnetic transition rates will be more strongly influenced than the electric transition rates because the average velocity of the mesons is higher than that of the nucleons and the magnetic transition rates depend on the current distribution while the electric rates depend on the charge distribution. For this last reason also, $E'_{l\mu}$ is in general much smaller than $E_{l\mu}$, and the electric transition rates are practically independent of the magnetic moments μ_j. The fact that the static magnetic moments of complex nuclei can be calculated, under the assumption of the additivity of the moments, with an accuracy of about 0.2 nuclear magnetons from sufficiently accurate wave functions indicates that the corrections to the transition rates may not be as large as might have been feared. This may be a less valid conclusion for the high-multipole moments than for the low ones. The selection rules are unaffected by the exchange moments.

12.3. Single Particle Matrix Elements

The expressions for the radiative transition rates can be evaluated most easily in the independent particle models for states which differ in the quantum numbers of a single nucleon. Under somewhat schematic assumptions for the radial dependence of the wave function of this single nucleon, one obtains:

$$\lambda(l, E) = \frac{18}{l+3} \frac{(l-1)!(l+1)!}{[(2l+1)!]^2} (2kR)^{2l} kc, \qquad (12.5a)$$

$$\lambda(l, M) = \frac{180}{l+3} \frac{(l-1)!(l+1)!}{[(2l+1)!]^2} (2kR)^{2l} \frac{e^2}{Mc^2 R} \frac{\hbar k}{MR}. \qquad (12.5b)$$

R is the nuclear radius. These rates are called Weisskopf units, and actual transition probabilities are conveniently expressed in terms of these units.

The expressions (12.5) are not the rates expected on the basis of the independent particle pictures; they are rates in terms of which the actual transition rates can be conveniently measured. Correction factors must be applied even if there is only a single particle outside of closed shells. These correction factors are less important for magnetic than for electric transitions because the neutron's and the proton's magnetic moments are of the same order of magnitude. They generate, therefore, magnetic fields of similar magnitude and (12.5b) give the corresponding transition rate.

If there is a single proton outside of closed shells, this proton and the rest of the nucleus will move about their common center of mass. Since the rest of the nucleus (the "core") also has a positive charge, and since proton and core are always on opposite sides of the center of mass, the dipole moment of the system will be decreased. This will lead to a decrease of the transition rate by a factor $(1 - Z/A)^2 = (N/A)^2$. This factor can be obtained directly from (12.4a) but follows more easily from the remark that two bodies, moving about their common center of mass and having the same charge-to-mass ratio, do not create any dipole field. If the extra particle outside of the closed shells is a neutron, it is the core which creates the dipole field and the transition rate is given by (12.5a) multiplied by $(Z/A)^2$. For higher electric multipole radiations the correction factor for the proton is negligible; that for the neutron is Z^2/A^{2l}, a very small number.

If there are several particles outside of closed shells, the transition rate will vanish under the assumption of the independent particle model if more than one of the orbits of initial and final states are different: the operators in (12.3) change the state of only one particle. There is a similar selection rule in atomic spectroscopy which is, in general, quite well obeyed. Even if initial and final configurations differ in only one orbit, the calculated rate will be, as a rule, well below one Weisskopf unit because the wave functions will contain many terms. This is apparent already from (7.1) and (7.2) even though these expressions apply for the case of only two particles. Each term of Ψ_i is connected by the operators of (12.3) with only those terms of Ψ_f in which all factors except one are also factors of the term in Ψ_i. This, and the possibility of cancellation of the terms—of which there still will be many—reduces the calculated transition rate below one Weisskopf unit unless there are phase relations between the contributions of the various terms.

The experimental material bears out these conclusions only partially. Dipole transitions, on the whole, conform to the rules given. They are usually only a fraction of a Weisskopf unit and their magnitude shows considerable spread—a factor of about five or ten in either direction from the average of about a tenth of a Weisskopf unit. There are several very weak transitions which *may* be interpreted as occurring between configurations which differ by more than one orbit. There is, however, no clear evidence for this last point.

Many of the higher multipole transition rates are surprisingly high. For E-2 transitions the collective model predicts such high rates and these were mentioned before as supporting the collective model. However, surprisingly high rates are observed also in regions for which the collective model is not usually considered to be appropriate. Thus, the E-2 transition from the 0.87 mev excited state of O^{17} to the normal state proceeds at the rate of one Weisskopf unit. There is only one particle outside of closed shells in this case and this is a neutron. The Z^2/A^{2l} factor is, therefore, less than 10^{-3} and the discrepancy is quite flagrant. On the whole, one gains the impression that the neutrons radiate about as strongly as the protons—just as they are responsible for quadrupole moments of similar magnitude. Another surprising fact is the lack of spread in the strengths of M-4 transitions, most of which proceed at the rate of about two Weisskopf units. As yet no fully satisfactory explanation has been found for these facts.

REFERENCES

1

The discovery of radioactivity is due to H. A. Becquerel (*Compt. Rend.* 122, 420, 501, 559, 689, 762, 1086 (1896); 123, 855 (1896)). The exciting story of the determination of the properties of radioactive substances and their radiations is also described in:

Mme. P. Curie, *L'isotopie et les Elements Isotopes*, Press Univ. Paris, 1924.

E. Rutherford, *Radioactive Substances and Their Radiations*. Cambridge Univ. Press, 1913.

K. Fajans, *Jahrb. d. Rad.* 14, 314 (1917).

It is difficult to single out a few articles which were principally responsible for the development of concepts of nuclear structure. Four articles of major importance are:

E. Rutherford, *Phil. Mag.* 21, 669 (1911)
H. Geiger and E. Marsden, *Phil. Mag.* 25, 610 (1913)
J. Chadwick, *Proc. Roy. Soc.* A136, 692 (1932)
W. Heisenberg, *Zeits. f. Physik* 77, 1 (1932).

The atmosphere in which Rutherford's model was discovered is recaptured by C. G. Darwin in the book, *Niels Bohr and the Development of Physics*, McGraw-Hill Book Co., New York, 1955.

The first measurements of nuclear masses are due to Aston and Dempster:

A. J. Dempster, *Phy. Rev.* 11, 316 (1918) and later articles
F. W. Aston, *Phil. Mag.* 38, 709 (1919) and later articles.

The modern techniques of mass spectroscopy are due to Bainbridge, Mattauch, Nier, Duckworth, and many others.

The relation between lifetime and decay energy of α-emitters was given by H. Geiger and J. M. Nuttall, *Phil. Mag.* 22, 613 (1911).

The explanation of this relation marks the beginning of modern nuclear theory. It was given simultaneously by:

G. Gamow, *Zeits. f. Physik*, 51, 204 (1928).
R. W. Gurney and E. U. Condon, *Nature* 122, 439 (1928), *Phys. Rev.* 33, 127 (1929).

Fission was discovered by Hahn and Strassmann:

O. Hahn and F. Strassmann, *Nature* 27, 11 (1939).

The nature and properties of γ- and β-radiations were uncovered principally by Ellis and Meitner. Two very important articles in this connection are:

L. Meitner, *Zeits. f. Physik*, 34, 807 (1925)
C. D. Ellis and W. A. Wooster, *Proc. Roy. Soc.* A117, 109 (1927).

The theory of β-decay is due to:

E. Fermi, *Zeits. f. Physik*, 88, 161 (1934),
cf. also H. A. Bethe and R. F. Bacher, *Rev. Mod. Phys.* 8, 82 (1936) Chapter VII.

The radioactivity of the neutron was first observed by:

A. H. Snell and L. C. Miller, *Phys. Rev.* 74. 1217 (1948).
cf. also J. M. Robson, *Phys. Rev.* 83, 349 (1951).

The lowest limit for the mass of the neutrino was obtained by D. R. Hamilton, W. P. Alford and L. Gross, *Phys. Rev.* 92, 1521 (1953).

Production of radioactive nuclei as a result of the interaction between neutrinos and stable neuclei was observed by:

F. Reines and C. L. Cowan, *Nature*, 178, 446 (1956);
C. L. Cowan, F. Reines, F. B. Harrison, H. W. Kruse, and A. D. McGuire, *Science* 124, 103 (1956).

The most careful measurement of the double β-decay is of recent origin:

M. Awschalom, *Phys. Rev.* 101, 1041 (1956);

cf. also R. C. Winter, *Phys. Rev.* 99, 88 (1955).
The explanation of isomerism is due to:
C. F. v. Weizsäcker, *Naturwiss.* 24, 813 (1936).

2

Early work on the systematics of isotopes is principally due to Aston:
F. W. Aston, *Mass Spectra and Isotopes*, E. Arnold and Co., London, 1933.
cf. also W. D. Harkins, *Phys. Rev.* 19, 136 (1922) and V. M. Goldschmidt, *Geochemische Verteilungsgesetze der Elements*, Norske Videnskape Akad. Oslo,1937.
The use of three binding energy surfaces was proposed by G. Gamow. The first semi-empirical mass formula was given by:
C. F. v. Weizsäcker, *Zeits. f. Physik*, 96, 431 (1935)
cf. also E. Feenberg, *Revs. Mod. Phys.* 19, 239 (1947)
The constants of the text are those of
M. H. L. Pryce, *Proc. Phys. Soc.* 63, 692 (1950).
The particular, stability of the nucleon numbers which are now called "magic" was noticed by:
W. Elsasser, *Jour. Phys. Rad.* 5, 389, 635 (1934).
Convincing evidence for the preferred nature of these numbers was presented by:
M. G. Mayer, *Phys. Rev.* 75, 1969 (1949).

3

First measurement of a nuclear magnetic moment, that of a proton, is due to:
O. Stern, I. Estermann and R. Frisch, *Zeits. f. Physik* 85, 4 (1933), 86, 132 (1933).
cf. also I. I. Rabi, J. M. B. Kellogg and J. R. Zacharias, *Phys. Rev.* 46, 157 (1934).
The article in which Schmidt's lines are defined is:
T. Schmidt, *Zeits. f. Physik* 106, 358 (1937).
The phenomenon of the quadrupole moment was discovered by:
H. Schüler and T. Schmidt, *Zeits. f. Physik* 94, 457 (1935).
The quadrupole moment of the deuteron was found by:
J. M. B. Kellogg, I. I. Rabi, N. F. Ramsey, and J. R. Zacharias, *Phys. Rev.* 56, 728 (1939): 57, 677 (1940).
For the measurement of the magnetic octupole moment:
Cf. V. Jaccarino, J. G. King, R. A. Satten, and H. H. Stoke, *Phys. Rev.* 94, 1798 (1954).
For measurement of nuclear moments in general see:
N. F. Ramsey, *Molecular Beams*, Clarendon Press, Oxford 1956, and *Nuclear Moments*, John Wiley and Sons, New York, 1953.
Determinations of the size of nuclei
(a) by scattering cross-section of fast neutrons:
R. Sherr, *Phys. Rev.* 68, 240 (1945).
E. Amaldi, D. Bocciarelli, C. Cacciaputo, G. Trabacchi, *Nuovo Cimento* 3, 203 (1946)
(b) by the lifetime—α-decay energy relation:
G. Gamow, *Constitution of Atomic Nuclei and Radioactivity*, Clarendon Press Oxford 1931, Chapter II.
(c) by the comparison of the binding energies of mirror nuclei:
E. P. Wigner and E. Feenberg, *Reports on Progress in Physics* 8, 274 (1941).
The energy levels of μ mesons, bound to atomic nuclei, were measured by V. L. Fitch and J. Rainwater, *Phys. Rev.* 92, 789 (1953). This method of obtaining the size of nuclei was suggested by J. A. Wheeler, *Revs. Mod. Phys.* 21, 133 (1949).
For the high-energy electron diffraction experiments, cf. R. Hofstadter, *Revs. Mod. Phys.* 28, 214 (1956), which also contains references to earlier literature.

4

The concepts of cross-section and differential cross-section originated in the kinetic theory of gases. See, for instance:

L. Boltzmann, *Vorlesungen über Gastheorie*, Johann Ambrosius Barth, Leipzig, 1898.

Much of the early work on nuclear theory was concerned with the effect of the electrostatic barrier on nuclear reactions. See, for instance:

G. Gamow, *Structure of Ayomic Nuclei and Nuclear Transformations*, Clarendon Press, Oxford, 1937;

also G. Breit, M. Ostrovsky and D. Johnson, *Phys. Rev.* 49 22 (1936).

Chapters IX and X of Gamow's book give an excellent summary of the ideas developed before 1937 by Gamow himself, by Bethe, Breit, Goldhaber, Landau, Mott, Oppenheimer, Peierls, Teller, Wick, and their collaborators, and many others. The book also reviews much of the experimental information avilable before 1937. Experimental work provided the main stimulus for introducing the concept of resonance reactions. This was done simultaneously by N. Bohr, *Nature* 137, 344 (1936) and by G. Breit and E. P. Wigner, *Phys. Rev.* 49, 519 (1936). cf. also the review article, *Amer. J. Phys.* 23, 371 (1955) which deals also with the limitations of the compound nucleus model.

The modern theory of stripping reactions is due to:

S. Butler, *Proc. Roy. Soc.* A202, 559 (1951).

The optical model was used in early theoretical work on nuclear reactions but was abandoned when found to be inadequate in the resonance region. The revival and proper appreciation of the range of applicability is the work of:

H. Feshbach, C. E. Porter and V. F. Weisskopf, *Phys. Rev.* 96, 448 (1954).

For its application to high energy processes:

S. Fernbach, R. Serber and T. B. Taylor, *Phys. Rev.* 75, 1352 (1949).

The phenomenon of Coulomb excitation was forseen by:

V. F. Weisskopf, *Phys. Rev.* 53, 1018 (1938).

It was found simultaneously, about fifteen years later, by:

T. Huus and C. Zupancic, *Dan. Mat. Fys. Medd.* 28, No. 1 (1953), and

C. McClelland and C. Goodman, *Phys. Rev.* 91, 760 (1953).

5

The approximate magnitude and range of the nuclear interaction was first determined by:

E. P. Wigner, *Phys. Rev.* 43, 252 (1933).

cf. also E. Feenberg and J. Knipp, *Phys. Rev.* 48, 960 (1935).

The effect of the tensor forces was investigated by:

W. Rarita and J. Schwinger, *Phys. Rev.* 59, 436 (1941).

L. Hulthén, *Arkiv. Nat. Astron. Fysik*, 35A, Section 25 (1948) gave a general review of our knowledge concerning the wave function of the deuteron.

The basic concepts of the meson theory of nuclear forces is due to H. Yukawa; Levy's calculation and its evalution, to:

M. M. Levy, *Phys. Rev.* 88, 725 (1952) and

A. Klein, *Phys. Rev.* 89, 1158 (1953).

The interaction (5.6) was proposed by:

H. H. Hall and J. L. Powell, *Phys. Rev.* 90, 912 (1953).

The significance of scattering experiments on ortho and parahydrogen was first recognized by:

J. S. Schwinger and E. Teller, *Phys. Rev.* 52, 286 (1937).

The idea of the repulsive core was first used by:

R. Jastrow, *Phys. Rev.* 81, 165 (1951).

The concept of saturation and of exchange forces is due to:

W. Heisenberg, *Zeits. f. Physik* 77, 1 (1932)

E. Majorana, *Zeits. f. Physik*, 82, 137 (1933).

The first of these articles already notes that the proton-proton and neutron-neutron interactions appear to be equal. Most careful interpretation of the proton-proton scattering experiments is due to G. Breit and his collaborators:

G. Breit, E. U. Condon, and R. D. Present, *Phys. Rev.* 50, 825 (1936);
G. Breit, H. M. Thaxton and L. Eisenbud, *Phys. Rev.* 55, 1018 (1939).

Equality of the proton-proton and proton-neutron interactions, in contrast to the equality of proton-proton and neutron-neutron interactions already noted by Heisenberg, was noted by:

B. Cassen and E. U. Condon, *Phys. Rev.* 50, 846 (1936), and by
G. Breit and E. Feenberg, *Phys. Rev.* 50, 850 (1936).

The isobaric spin quantum number was introduced by:

E. P. Wigner, *Phys. Rev.* 51, 106 (1937).

Effect of the electrostatic interaction on the validity of the quantum number T was investigated by:

L. A. Radicati, *Proc. Phys. Soc.* A66, 139 (1953); A67, 39 (1954), and
W. M. MacDonald, *Phys. Rev.* 98, 60 (1955); 100, 51 (1955); 101, 271 (1956).

For experimental information pertaining to isobaric spin, cf. the series of articles by Wilkinson and his collaborators, beginning with:

D. H. Wilkinson and G. A. Jones, *Phil. Mag.* 44, 542 (1953).
See also D. H. Wilkinson, *ibid*, 1, 1031 (1956).

6

The constrasting names "Powder Model" and "Shell Model" are due to M. H. L. Pryce. The most brilliant description of the powder model is due to:

N. Bohr, *Nature* 137, 344 (1936).

The considerations of the text were given by E. P. Wigner, *Phys. Rev.* 51, 947 (1937) and W. H. Barkas, *Phys. Rev.* 55, 691 (1939).

cf. also E. Feenberg and E. P. Wigner, *Reports on Progress in Physics* 8, p. 274 (1941).

7

Early attempts to interpret nuclear structure on the basis of the independent particle model:

G. Beck, *Zeits. f. Physik*, 61, 615 (1930).
G. Gamow, *Zeits. f. Physik* 89, 592 (1934)
K. Guggenheimer, *J. Phys. et Rad.* 5, 475 (1934)
W. Elsasser, *J. Phys. et Rad.* 5, 635 (1934)
F. Hund, *Phys. Zeits.* 38, 929 (1937).

The method for calculating properties of "nuclear matter" has been given by:

K. A. Brueckner and C. A. Levinson, *Phys. Rev.* 97, 1344 (1955)

It was described also by:

H. A. Bethe, *Phys. Rev.* 103, 1353 (1956), J. Goldstone, *Proc. Roy. Soc.* A239, 267 (1957).

It is based on the scattering theory of:

K. M. Watson, *Phys. Rev.* 89, 575 (1953).

The underlying ideas were fully developed and brought into a truly convincing form by:

N. M. Hugenholtz, Physica 23, 533 (1957).

The fact that the $1p$ shell is being completed in the region between $A = 4$ and $A = 16$ was recognized by:

J. H. Bartlett, *Nature* 130, 165 (1932).

Calculations on the L-S model were carried out by:

E. Feenberg and E. P. Wigner, *Phys. Rev.* 51, 95 (1937),
M. E. Rose and H. A. Bethe, *Phys. Rev.* 51, 205 (1937).

The j-j shell model was proposed by:

O. Haxel, J. H. D. Jensen and H. E. Suess, *Zeits. f. Physik* 128, 295 (1950),
M. G. Mayer, *Phys. Rev.* 75, 1969 (1949); 78, 16 and 22 (1950); 79, 1012 (1950).
Calculation of the fraction of the wave function which belongs to the lowest supermultiplet is contained in the article:
A. R. Edmonds and B. H. Flowers, *Proc. Roy. Soc.* A229, 536 (1955).
This is one of a series of articles, by Edmonds, Elliot, Flowers, and Lane, on the transition from L-S to j-j coupling.
See also M. G. Redlich, *Phys. Rev.* 99, 1427 (1955). D. Kurath, ibid. 101, 216 (1956).

The problem was first broached by:
D. R. Inglis, *Revs. Mod. Phys.* 25, 390 (1953),
who first recognized the transitional nature of the states of light nuclei.

It is not possible to give an adequate review of the rapidly growing literature on the j-j model. Much of it is reviewed in the monographs:
M. G. Mayer and J. H. D. Jensen, *Elementary Theory of Nuclear Shell Structures*, John Wiley and Sons, New York, 1955, and
E. Feenberg, *Shell Theory of the Nucleus*, Princeton University Press, 1955.
Table 7.1 is based on several articles and reviews, including:
P. F. A. Klinkenberg, *Revs. Mod. Phys.* 24, 63 (1952).
N. Zeldes, *Nuclear Physics* 2, 1 (1956).
The coupling rules for even-even, even-odd, and odd-even nuclei are contained already in the two articles of Mayer which were mentioned last. The coupling rules for odd-odd nuclei were formulated by:
L. W. Nordheim, *Phys. Rev.* 78, 294 (1950).
An explanation for these in terms of nuclear forces was proposed by:
C. Schwartz, *Phys. Rev.* 94, 95 (1954).
cf. also A. de Shalit, *Phys. Rev.* 91, 1479 (1953), D. Kurath, *ibid*. 91, 1430 (1953).
An apperciable fraction of the data contained in Table 7.2 was originally obtained by:
M. Goldhaber and R. D. Hill, *Revs. Mod. Phys.* 24, 179 (1952)
M. Goldhaber and A. W. Sunyar, *Phys. Rev.* 83, 906 (1951).
The table itself is based on the data collection of Way, King, McGinnis, and van Lieshout with a few additions.

The theories of the magnetic moments are reviewed by:
R. J. Blin-Stoyle, *Revs. Mod. Phys.* 28, 75 (1956).
Calculations on the effect of configuration interaction on magnetic and quadrupole moments are due to:
H. Horie and A. Arima, *Prog. Theo. Phys.* 11, 509 (1954); *Phys. Rev.* 99, 778 (1955).
The last few paragraphs of the text are based on the papers of:
S. Goldstein and I. Talmi, *Phys. Rev.* 102, 589 (1956) and
I. Talmi and R. Thieberger, *Phys. Rev.* 103, 718 (1956).
A partial explanation of the spin-orbit coupling, postulated by the j-j shell model, was given by A. M. Feingold, *Phys. Rev.* 101, 258 (1956); 105, 944 (1957). See also J. Keilson, ibid. 82, 759 (1951).

8

The most significant papers on the α-particle model are:
W. Wefelmeier, *Naturw.* 25, 525 (1937)
J. A. Wheeler, *Phys. Rev.* 52, 1083 (1937)
L. R. Hafstad and E. Teller, *Phys. Rev.* 54, 681 (1938)
B. O. Grönblom and R. E. Marshak, *Phys. Rev.* 55, 229 (1939)
D. M. Dennison, *Phys. Rev.* 57, 454 (1940); 96, 378 (1954),
H. Margenau, *Phys. Rev.* 59, 37 (1941).
A. E. Glassgold and A. Galonsky, *Phys. Rev.* 103, 701 (1956).
Observation of the easy deformability of the originally spherical nuclear matter was made by:
J. Rainwater, *Phys. Rev.* 79, 432 (1950).

Starting from this observation, the collective model was developed in a series of articles by A. Bohr, Mottelson and their collaborators, beginning with A. Bohr and B. R. Mottelson, *Physica* 18, 1066 (1952) and by D. L. Hill and J. A. Wheeler, *Phys. Rev.* 89, 1102 (1953).

Attempts to show that the collective model is in conformity with quantum mechanics and our knowledge of nuclear forces are too numerous to list fully. Two representative papers are:

H. A. Tolhoek, *Physica* 21, 1 (1955),
H. Lipkin, A. de Shalit and I. Talmi, *Nuovo Cim.* 2, 773 (1955).

The rotational spectra were postulated by:

N. Bohr and F. Kalckar, *Dan. Mat. Fys. Medd.* 14, No. 10 (1937), and
H. A. Bethe, *Revs. Mod. Phys.* 9, 69 (1937).

However, the modern theory of the rotational spectra is due to:

A. Bohr and B. R. Mottelson, *Dan. Mat. Fys. Medd.* 27, No. 16, (1953); *Phys. Rev.* 90, 717 (1953)
G. Alaga, K. Alder, A. Bohr and B. R. Mottelson, *ibid.* 29, No. 9 (1955).

The existence of two regions in which the collective model has particular validity was first noted by:

G. Scharff-Goldhaber, *Phys. Rev.* 103, 837 (1956).
cf. also G. Scharff-Goldhaber and J. Weneser, *ibid.* 98, 212 (1955) and
K. Way, D. N. Kundu, C. L. McGinnis and R. van Lieshout, *Annual Reviews of Nuclear Science* 6, Annual Reviews, Inc, Stanford, California, 1956.

K. W. Ford, *Phys. Rev.* 95, 1250 (1954) was the first to draw attention to the discrepancy between the magnitudes of nuclear deformations as determined from quadrupole moments and from rotational spectra. See also:

A. W. Sunyar, *Phys. Rev.* 98, 653 (1955).

Attempts to reconcile the two sets of data were initiated by:

D. R. Inglis, *Phys. Rev.* 96, 1059 (1954).

See also A. Bohr and B. R. Mottelson, *Dan. Mat. Fys. Medd.* 30, No. 1 (1955).

9

The best known book on scattering theory is N. F. Mott and H. S. W. Massey, *The Theory of Atomic Collisions*, Clarendon Press, Oxfrod, 1949. Only a few of the of the most important contributions to the theory of nuclear collisions will be named.

M. Born, *Zeits. f. Physik* 37, 803 (1926) is the originator of collision theory and of the "Born approximation". A generalization of this mathod is due to:

P. A. M. Dirac, *The Principles of Quantum Mechanics*, The Clarendon Press, Oxford, 1947, Chapter VIII.

H. Faxén and J. Holtsmark, *Zeits. f. Physik* 45, 307 (1927) introduced the method of decomposing plane waves into spherical waves, with definite angular momenta.

cf. also J. W. S. Rayleigh, *The Theory of Sound*, Macmillan and Co., London, 1926. Vol. 2, 334ff.

Exact theory of scattering by an electric point charge was first given by W. Gordon, *Zeits. f. Physik* 48, 180 (1928).

The modern theory of proton-neutron scattering is due, in addition to Breit and his collaborators, principally to:

L. Landau and J. Smorodinsky, *J. Phys. USSR* 8, 154 (1944)
J. S. Schwinger, *Phys. Rev.* 72, 742 (1947)
J. M. Blatt and J. D. Jackson, *Phys. Rev.* 76, 18 (1949).

The variational principle was introduced by:

L. Hulthén, *K. Fys. Selsk. Lund. Ferh.* 14, 8, 21 (1944); *Arkiv. Mat. Astr. Fys.* 35, No. 25 (1948).
B. A. Lippmann and J. S. Schwinger, *Phys. Rev.* 79, 469 (1950).

Very high-energy nuclear collisions are treated, for instance, by:

R. Serber, *Phys. Rev.* 72, 1114 (1947)
G. F. Chew and M. L. Goldberger, *Phys. Rev.* 73, 1409 (1948)
S. Fernbach, R. Serber and T. B. Taylor, *Phys. Rev.* 75, 1952 (1949).

The impulse approximation is given in:

REFERENCES

G. F. Chew, *Phys. Rev.* 80, 196 (1950)
G. F. Chew and G. C. Wick, *Phys. Rev.* 85, 636 (1952).
The concept of the collision matrix is due to:
J. A. Wheeler, *Phys. Rev.* 52, 1107 (1937).
See also P. L. Kapur and R. Peierls, *Proc. Roy. Soc.* A166, 277 (1938)
W. Heisenberg, *Zeits. f. Physik* 120, 513, 673 (1943).
Extension of the formulæ (9.3) to take the electrostatic interaction into account is principally due to G. Breit and to his collaborators. The picture of the compound nucleus was given simultaneously by:
N. Bohr, *Nature* 137, 344 (1936), and
E. P. Wigner and G. Breit, *Bull. Amer. Phys. Soc.* Feb. 1936,
G. Breit and E. P. Wigner, *Phys. Rev.* 49, 519 (1936).
The last article gives the resonance formula.
The derivation of the resonance formula as given in the text utilizes much earlier work.
It follows the article:
E. P. Wigner and L. Eisenbud, *Phys. Rev.* 72, 29 (1947).
cf. also J. M. Blatt and V. F. Weisskopf, *Theorical Nuclear Physics*, John Wiley and Sons, New York, 1952, Chapters VIII and X
R. G. Sachs, *Nuclear Theory*, Addison Wesley Press, Cambridge 1953, pp. 290–304.
These two books give an excellent survey of the knowledge of nuclear reactions (and also of nuclear structure) up to 1951 or 1952.
For the effect of the electrostatic interaction, see:
G. Breit and J. S. McIntosh, *Phys. Rev.* 106, 246 (1957)
For the first derivations of (9.13), see:
H. A. Bethe, *Revs. Mod. Phys.* 9, 69 (1937)
H. Feshbach, D. C. Peaslee, V. F. Weisskopf, *Phys. Rev.* 71, 145 (1947).
The concept of the "radioactive state" is principally due to G. Gamow. Cf. the reference to Chapter 4.
Detailed development of the strong coupling theory, in the form of the statistical theory, is due to:
V. F. Weisskopf and D. H. Ewing, *Phys. Rev.* 57, 472, 935 (1940).
See also the book of Blatt and Weisskopf, quoted above, Chapter IX.
The experimental information on the average cross-sections was obtained chiefly by Barschall and his collaborators. See:
M. Walt and H. H. Barschall, *Phys. Rev.* 93, 1062 (1954).
Also R. K. Adair, *Revs. Mod. Phys.* 22, 249 (1950).
Barschall found the maxima in the average cross-section curve which are often referred to as Barschall maxima. A maximum, predicted by the cloudy crystal ball model, which was not known before, was found by:
R. Cote and L. M. Bollinger, *Phys. Rev.* 98, 1162 (1955).
The cloudy crystal ball model was proposed by:
H. Feshbach, C. Porter and V. F. Weisskopf, *Phys. Rev.* 96, 448 (1954).
See also R. K. Adair, *ibid*, 94, 737 (1954)
R. D. Woods and D. S. Saxon, *ibid.* 95, 377 (1954).
The intermediate coupling model was proposed simultaneously by:
J. M. C. Scott, *Phil. Mag.* 45, 1322 (1954) and
E. P. Wigner, *Science*, 120, 790 (1954).
Its detailed development is however due to:
A. M. Lane, R. G. Thomas and E. P. Wigner, *Phys. Rev.* 98, 693 (1955)
E. Vogt, *Phys. Rev.* 101, 1792 (1956)
C. Bloch, *Nuclear Physics.* 4, 503 (1957)

10

For Born's collision theory, see notes to chapter 9.
The theory of the stripping reactions was given by:
S. T. Butler, *Proc. Roy. Soc.* A208, 559 (1951).

Similar results were obtained, independently, by F. Friedman and W. Tobocman (unpublished). The importance of (d, p) and (d, n) reactions was recognized by J. R. Oppenheimer and M. Phillips, *Phys. Rev.* 48, 500 (1935). Butler's results were obtained on the basis of Born's approximation by:

A. B. Bathia, Kun Huang, R. Huby and H. C. Newns, *Phil. Mag.* 43, 485 (1952)
P. B. Daitch and J. A. French, *Phys. Rev.* 87, 900 (1952).

See also H. McManus and W. T. Sharp, *Phys. Rev.* 87, 188 (1952); N. Austern, S. T. Butler and H. McManus, ibid. 92, 350 (1953) for an extension of the underlying concepts.

For the early papers on Coulomb excitation, see the notes to Chapter 4. Accurate calculation of the dipole transitions was carried out by:

R. Huby and H. C. Newns, *Proc. Phys. Soc.* A64, 619 (1951) and
C. T. Mullin and E. Guth, *Phys. Rev.* 82, 141 (1951).

The significance of the field was recognized and expressions for all multipole transsitions were obtained by:

K. A. Ter Martirosyan, *J. Exptl. Theor. Phys. USSR* 22, 284 (1952)
A. Bohr and B. R. Mottelson, *Dan. Mat. Fys. Medd.* 27, No. 3 (1953)
K. Alder and A. Winther, *Phys. Rev.* 91, 1578 (1953); 96, 237 (1954)
F. D. Benedict, P. B. Daitch and G. Breit, *ibid*, 101. 171 (1956).

11

For the foundation of the theory, see the references to Section 1.3.

G. Gamow and E. Teller, *Phys. Rev.* 49, 895 (1936)

were the first to suggest modifications of the form of interaction proposed by Fermi. The five invariant forms were given by:

H. A. Bethe and R. F. Bacher, *Revs. Mod. Phys.* 8, 82 (1936).

This article summarizes the thinking on nuclear structure and nuclear reactions through 1935 in a most admirable fashion. H. Yukawa and S. Sakata, *Proc. Phys. Math. Soc.* Japan 17, 467 (1935) extended Fermi's theory for the process of electron capture.

T. D. Lee and C. N. Yang, *Phys. Rev.* 104, 254 (1956); 105, 1671 (1957),
T. D. Lee, R. Oehme and C. N. Yang, *Phys. Rev.* 107, 340 (1957)

Recent considerations of the above have been confirmed experimentally on the β-decay of Co^{60} by:

E. Ambler, R. W. Hayward, D. D. Hoppes and R. P. Hudson, *Phys. Rev.* 105, 1413 (1957).

They show that the neutrino is more appropriately described by a two-component wave function. These ideas have important bearing on the theory of β-decay which, however, do not seem to affect those parts of the theory which are presented in the text. They do imply, however, that the parity law is violated in β transitions. The interaction is the sum of an expression (11.2), as explained in the text, and of another expression with the same constants g_K in which the scalar formed from the proton and neutron operators is multiplied with the pseudoscalar formed from the electron and neutrino operators; the vector formed from the proton and neutron operators is multiplied with the axial vector formed from the electron and neutrino operators, and so on. The terms of the expression (11.2) are consistent with the parity postulate (they were derived on the basis of that postulate); the terms which are to be added violate the parity rule.

The form of the interaction, given in the text, is the result of a long exploration. The existence of at least two types of interactions follows most conclusively from the observations of:

R. Sherr and J. B. Gerhart, *Phys. Rev.* 91, 909 (1953).

Even before, H. Mahmoud and E. Konopinski, *Phys. Rev.* 88, 1266 (1952) adduced evidence that the interaction consists of a tensor, a scalar, and a pseudoscalar part. This was confirmed by investigations on the angular correlations between the emitted electrons and neutrinos referred to later. We are indebted to Dr. J. B. Gerhart for a discussion of the present information on the magnitude of the interaction constants.

The transitions of various degrees of forbiddenness were classified by:

G. Uhlenbeck and E. J. Konopinski, *Phys. Rev.* 60, 308 (1941).
See also R. Marshak, *ibid* 61, 431 (1942), and
E. J. Konopinski, *Revs. Mod. Phys.* 15, 209 (1943) and his article in *Beta and Gamma Ray Spectroscopy* (edited by K. Siegbahn), North Holland Publishing Co., Amsterdam 1955.

The shape of the allowed spectra was calculated by Fermi. Early experiments gave conflicting results for the shape; the discrepancies were cleared up principally by:
A. W. Tyler, *Phys. Rev.* 56, 125 (1939) and
J. L. Lawson, *ibid.* 56, 131 (1939).

For the accuracy with which the spectrum of the electrons agrees with theory, see C. S. Wu's article in *Beta and Gamma Ray Spectroscopy*, quoted before. The measurements are due principally to herself, and to L. M. Langer, D. R. Hamilton and collaborators, and now extend from the upper limit down to about 10 kev.

Forbidden shapes were definitely identified first by:
L. M. Langer and H. C. Price, *Phys. Rev.* 75, 1109; 76, 641 (1949)
C. L. Peacock and A. C. G. Mitchell, *ibid.* 75, 1273 (1949)
C. S. Wu and L. Feldman, *ibid.* 76, 698 (1949).

M. Fierz, *Zeits. f. Physik* 104, 533 (1937) noticed the existence of interference terms in the spectral distribution of electrons from allowed β-transitions. The importance of the correlations between the directions in which electron and neutrino are emitted was pointed out by:
D. R. Hamilton, *Phys. Rev.* 71, 456 (1947).
B. M. Rustad and S. L. Ruby, *Phys. Rev.* 89, 880 (1953) and
J. L. Allen and W. K. Jentschke, *ibid.* 89, 902 (1953)
adduced evidence for the existence of the tensor interaction, and
W. P. Alford and D. R. Hamilton, *ibid.* 95, 1351 (1954) and
D. R. Maxson, J. S. Allen and W. K. Jentschke, *ibid.* 97, 109 (1955)
that of the scalar interaction by means of these angular correlations.

For the distinction between favored and unfavored allowed transitions, see:
L. W. Nordheim and F. L. Yost, *Phys. Rev.* 51, 943 (1937)
E. P. Wigner, *ibid.* 56, 519 (1939)
M. Bolsterli and E. Feenberg, *ibid.* 97, 736 (1955).
R. W. King, *ibid.* 99, 67 (1955).

For analysis of the experimental material on favored transitions, see:
E. Feenberg, *Phys. Rev.* 97, 736 (1955).

12

Methods of detection of γ-quanta are discussed, for instance, in *Beta and Gamma Ray Spectroscopy*, edited by K. Siegbahn, Chapters IV-VII. North Holland Publishing Co., Amsterdam, 1955. The coefficient of internal conversion plays an important role in determination of the multipolarity of radiation. Early work of H. R. Hulme and of his collaborators is summarized in Gamow's book (see notes to Chapter 4). The most elaborate calculations are due to M. E. Rose. See Appendix IV of *Beta and Gamma Ray Spectroscopy*, quoted above. For the effect of the finite size of the nucleus, see:
L. A. Sliv, *J. Exp. Theor. Phys.* 21, 77 (1951)
L. A. Sliv and M. A. Listengarten, *ibid.* 22, 29 (1952).
Also F. K. McGowan and P. H. Stelson, *Phys. Rev.* 103, 133 (1956).

Pair production in O^{16} was first treated by:
J. R. Oppenheimer, *Phys. Rev.* 60, 164 (1941).

For a more complete discussion, see:
R. H. Dalitz, *Proc. Roy. Soc.* A206, 521 (1951).

The quantum theory of radiation was founded by:
P. A. M. Dirac, *Proc. Roy. Soc.* A114, 243, 710 (1927).

However, the theory has been greatly generalized since. The best known comprehensive treatment is W. Heitler's, *The Quantum Theory of Radiation*, Clarendon Press, Oxford, 1947. The similarity between many details of Dirac's paper (e.g. choice of gauge) and the most modern treatments remains interesting. Steps leading to (12.2) are given in detail in Blatt and Weisskopf's *Theoretical Nuclear Physics*, Chapter XII, John Wiley and Sons, New York, 1952.

REFERENCES

The problem of exchange currents was raised by:
E. U. Condon and G. Breit, *Phys. Rev.* 52, 787 (1937).
See also C. Møller and L. Rosenfeld, *Dan. Mat. Fys. Medd.* 20, No. 12 (1943).
These articles are based on the field theory of nuclear forces. The phenomenological approach was proposed by:
R. G. Sachs and N. Austern, *Phys. Rev.* 81, 705 (1951).
See also R. K. Osborn and L. L. Foldy, *ibid.* 79, 795 (1951)
F. Villars, *ibid.* 86, 476 (1952) and
R. G. Sachs, *Nuclear Theory*, Chapter 9, Addison Wesley Publishing Co., Cambridge, 1953, which gives a comprehensive account of the phenomenological point of view.
The units (12.5) were established by:
V. F. Weisskopf, *Phys. Rev.* 83, 1073 (1951).
See also J. M. Blatt and V. F. Weisskopf, *Theoretical Nuclear Physics*, quoted above.
The review of the experimental material is a very inadequate summary of several articles. These include:
D. R. Inglis, *Revs. Mod. Phys.* 25, 390 (1053)
A. M. Lane, *Proc. Roy. Soc.* A68, 189, 197 (1955)
M. Goldhaber and A. W. Sunyar, *Beta and Gamma Ray Spectroscopy*, quoted above and
D. H. Wilkinson, *Phil. Mag.* 1, 127 (1956).

SOME RECENT COLLECTIONS OF NUCLEAR DATA.

A more complete list is being published by R. W. Gibbs of the National Research Council.

1. J. Stehn. General Electric Company, *Chart of Nuclides*. 5th edition, 1956.
2. A. H. Wapstra. Isotopic Masses. *Physica* 21, 367, 385 (1955).
3. I. W. Hay, Atomic Masses. *Atomic Energy of Canada Ltd, Report TPI* 78 (1955).
4. H. E. Suess and H. C. Urey. Abundances of the Elements. *Revs. Mod. Phys.* 28, 53 (1956).
5. E. F. Kjul'c, V. V. Kunc and V. G. Hartman. Table of Nuclear Moments *Usp. Fiz. Nauk* 55, 537 (1955).
6. J. R. Blin-Stoyle. Theories of Nuclear Moments. *Revs. Mod. Phys.* 28, 75 (1956).
7. N. F. Ramsey. *Nuclear Moments*, John Wiley and Sons, New York, 1953.
8. F. Ajzenberg and T. Lauritsen. Energy Levels of Light Nuclei V. *Revs. Mod. Phys.* 27, 77 (1955).
9. P. M. Endt and J. C. Kluyver. Energy Levels of Light Nuclei. *Rev. Mod. Phys.* 26, 95 (1954).
10. K. Way, R. W. King, C. L. McGinnis and R. van Lieshout. Nuclear Level Schemes. *U. S. Atomic Energy Commission Report TID-5300* (1955).
11. K. Way, L. Fano, M. R. Scott and K. Thew. Nuclear Data. *National Bureau of Standards Circular 499*, with 3 supplements. Current work is being reviewed in special issues of *Nuclear Science Abstracts*. This collection is probably the most comprehensive though by now somewhat difficult to use. It is arranged by nuclides rather than by properties. It gives abundances, spins and moments, lifetimes, modes of disintegration. It does not give nuclear masses.
12. R. W. King, N. M. Dismuke and K. Way. Tables of log ft values. *Oak Ridge National Laboratory Report 1450* (1952).
13. W. T. Sharp, H. E. Gove and E. B. Paul. Graphs of Coulomb Functions. *Atomic Energy of Canada Ltd. report 268* (1955).
14. H. Feshbach and M. M. Shapiro. Penetrabilities for Charged Particle Reactions. *U. S. Atomic Energy Commission, New York Operations Office, report 3077* (1953).
15. AEC Neutron Cross Section Advisory Group. Neutron Cross Sections. *Brookhaven National Laboratory Report 2040* (1955).
16. E. Vogt. The Widths and Spacings of Nuclear Resonance Lines. *U. S. Atomic Energy Commission, New York Operations Office, report NDA 14* (1955).
17. B. C. Dzelepov and L. K. Peker (Akademii Nauk CCCR, Moskva 1957) give diagrams for the excited states of all nuclei.

INDEX

INDEX

α-particle model, 65
α-radioactivity, 7, 8
  theory of, 81
Angular distributions, 74, 89
  in stripping processes, 92, 93
Angular correlations in electromagnetic transitions, 105
Angular momentum, 17, 41 (See spin-, orbital-angular momentum)
Anomalous quadrupole moments, 19

Barn, 24
Barrier penetration, 7, 23, 28 (See electrostatic barrier)
β-decay, 9f, 97ff
  allowed transitions, 100, 101, 103
  degree of forbiddenness, 102
  double β-decay, 10, 11
  favored transitions, 45, 49, 103
  forbidden transitions, 100f
  Gamow-Teller interactions, 98
  interaction Hamiltonian, 97
  lifetimes, 9, 10, 103
  nuclear matrix elements, 99, 103
  selection rules, 100, 101
  shape of spectrum, 101, 102
  theory, 97ff
  transition rates, 99
Bethe, 87
Binding energy, 4, 5
  discontinuities at magic nuclei, 47
Binding fraction, 5
Bohr, 84
Born approximation, 90
Breit, 79
Breuckner, 61
Butler, 90, 94

Center of mass coordinates, 24, 25
Chadwick, 3
Charge independence of nuclear forces, 35, 36f
Clebsch-Gordan coefficients, 52
Clouded crystal ball model, 83f
Collective coordinates, 67
Collective model, 65ff
  anomalous quadrupole moments, 65f
  comparison with j-j model, 68
  electric transition rates, 68, 96, 111
  rotational states, 67, 68
Collision matrix, 73f, 78f
Compound nucleus, 26f, 75

Configuration interaction, 60, 69
Configuration space, 71
Configuration splitting in j-j model, 56
Coulomb excitation, 29, 94 (See electric excitation)
Coupling rules in j-j model, 55ff
Cross-sections, 24, 73f, 75f
  average cross-sections, 84, 87

Decay probability, 6, 9
Deformation of nuclear matter, 66
Degree of forbiddenness, 102. (See β-decay forbidden transitions)
Density of nuclear matter, 21, 23
Deuteron properties, 32
Differential cross-section, 24
Diffuseness of nuclear surface, 22, 23
Dipole moments, see magnetic dipole moments
Dirac matrices, 99
Direct processes, 28, 29 (See surface reactions)
Double β-decay, 10, 11

Einstein, 4
Electric excitation, 29, 94f
Electric multipole moments, 17
Electric multipole operators, 95, 106f
Electric multipole radiation, angular momentum and parity of, 106
Electromagnetic transitions, 8, 105ff
  lifetimes, 8
  selection rules, 108, 110
Electromagnetic transition rates, 107
  collective model, 68, 96, 111
  independent particle model, 109, 110
Electron-nucleon interaction, 22
Electron scattering, high energy, 22
Electrostatic barrier
  α-decay, 7
  nuclear collisions, 28, 29, 30
Electrostatic energy, 21, 42
Even-even, even-odd, odd-odd nuclei, 12, 14, 44f
Exchange currents, 109
Excitation function, 26
Exclusion principle, 35, 48

Fermi, 9, 24, 97, 98, 102
Fission, 8
ft values, 103 (See β decay lifetimes)
γ-radiation, 8 (See electromagnetic transitions)

INDEX

Gamow, 98
Giant resonance model, 84 (See intermediate coupling model)

Half-life, 6
Hamiltonian
   β-decay, 97
   electric excitation, 94
Heisenberg, 3, 6
Heisenberg uncertainty relation, 6
Hofstadter, 22

Independent particle models, 41, 42, 47ff (See shell models)
   electromagnetic transition rates, 109
   selection rules, electromagnetic transitions, 110
Interactions of nucleons, see nuclear interactions
Interconfigurational interaction, 63
Intermediate coupling model, 84ff
   reduced widths in, 86
   relation to clouded crystal ball model, 87f
Internal conversion, 105, 106
Isobaric nuclei, 10, 12, 14
Isobaric spin, see isotopic spin
Isobars, 3
Isomers, 8, 108
Isotones, 3
Isotopes, 3
Isotopic spin, 36ff, 42f
   selection rules for, 39
Iostopic spin multiplets, 37, 38

j-j coupling model, 45, 48, 50f, 52ff
   β-decay transition rates, 61
   coupling rules, 55f, 59
   magic numbers, 54
   magnetic moments, 56
   parity assignments, 57
   parity of shells, 55
   problems of, 60f
   quadrupole moments, 60
   quantum numbers of low states, 57f
   quantum numbers of nucleons, 48
   radiative transition rates, 110
   spin orbit splitting, 54
   subshells, table of, 54

K-capture, 9

Level separations, 27
Level widths, 6, 27, 28, 75f, 78, 87
Levy interaction, 33
Lifetimes of decay processes, 6ff (See α-decay, β-decay, electromagnetic transitions)
L-S coupling shell model, 41, 48, 49ff

j-j model, comparison, 50f
   light nuclei, application to, 49f
   quantum numbers of nucleons, 48
   spin dependent forces, 50
   supermultiplet theory, relation to, 49

Magic nuclei, 47, 54
Magic numbers, 15, 16, 47
   relation to j-j model, 47
Magnetic dipole moments, 17f
   in j-j model, 56, 59
   of nucleons, 2, 59, 109
   Schmidt lines, 18
Magnetic multipole radiation, 106f
Majorana exchange forces, 33, 42, 44, 45
Majorana exchange operator, 33
Many particle models, 65ff (See α-particle model, collective model)
Mass energy relation, 4
Masses of nuclei, 4 (See binding energies)
Masses of nucleons, 4
Mass number, 3
Mass surfaces, 14
Matrix elements
   β-decay transitions, 99
   electromagnetic transitions, 105, 107, 108, 109
Mesic atoms, 21
Meson nucleon interaction, 22
Mirror nuclei, 20, 38
Models for nuclear reactions
   clouded crystal ball model, 82f
   intermediate coupling model, 84f
   optical model, 28, 82f
   strong coupling model, 83
Models of nuclei
   independent particle models, 47ff
   many particle models, 65ff
   powder models, 41f

Neutrino, 9, 10
   and double β-decay, 10
Neutrino rest mass, 9, 11, 102
Neutrino spin, 9, 11
Neutron half life, 10
Neutron-neutron interaction, 23, 32
Neutron-proton interaction, 32ff
Neutron proton scattering, 32
Nuclear interactions, 32ff
   charge independence, 36f
   Levy interaction, 33
   range, 32
   repulsive core, 34
   spin dependence of, 32
   tensor interaction, 33
Nuclear instability, 6ff (See α-decay β-decay fission, γ-radiation, particle emission)

INDEX

Nuclear matter, 40
Nuclear models, see models of nuclei
Nuclear reactions, 24ff, 71ff, 89ff
   direct processes, 28, 89ff
   resonance processes, 26, 75, 77
   surface reactions, 89ff
   survey of, 24ff
   table of common reactions, 31
Nuclear surface, diffuseness of, 22f
Nucleons, 3
Nucleon magnetic moments within nuclei, 59, 109
Nucleon properties, table of, 4

Odd-odd nuclei, 12, 14, 15, 17
   in supermultiplet theory, 44, 46
Optical model, 28, see clouded crystal ball model
Orbital angular momentum, 45, 49, 50

Pair production in nuclear transitions, 106
Pairing energy, 56
Parity
   assignments in j-j model, 57
   of multipole radiation, 106
   of nuclear states, 41
   selection rules in β-decay, 101
   selection rules in electromagnetic radiation, 108
   of shells in j-j coupling, 55
Partial widths, 27, 78f
Particle emission, 8, 27
Pauli, 9
Penetration factors, 79
Phase shift, 73
Pick up processes, 29, 90
Powder model, 41ff
Proton-proton interaction, 35
Proton-proton scattering, 35

Q of reaction, 26
Quadrupole moments, 19, 20, 65, 66
   in collective model, 67f
   in j-j model, 60
   of deuteron, 32, 33
Quasi-stationary state, 27

Radii of nuclei, 20ff
Radioactivity
   α-decay, 7f, 81
   β-decay, 9f, 97ff
Range of nuclear interactions, 32
Reactions, table of, 31
Reciprocity theorem, 74
Redlich, 56
Reduced widths, 79, 81, 86, 87

Relativistic invariance in β-interaction, 98
Resonance levels, 27, 75
Resonance reactions, 26, 27, 75, 77
Resonance scattering of radiation, 105
R-matrix, 77ff
Rotational states in collective model, 67, 68
Russell-Saunders coupling in atoms, 50
Rutherford, 3
Rutherford scattering, 95

Saturation properties, 35f
Schiff, 22
Schmidt lines, 18
Segré chart, 4, 13
Selection rules
   in β-decay, 100, 101
   in electromagnetic transitions, 108, 110
   for isotopic spin, 39
Separation energies, 5
Shape of nuclear surface, 66
Shell models, 41ff (See j-j coupling model, L-S coupling model)
Shell model for atoms, 41
Size of nuclei (See nuclear radii)
Spherical wave analysis
   in collisions, 72, 73
   in radiative transitions, 106
Spin, 17, 41
   in j-j model, 55, 57
   in L-S model, 50
   of normal states of nuclei, 17
   of nucleons, 2
Spin angular momentum, 36, 37
Spin dependence of nuclear forces, 32, 33
Spin orbit splitting, 54, 61
Statistics of nucleons, 2
Stripping reactions, 29, 90f
Strong coupling model, 83
Supermultiplet theory, 42ff
   β-transition rates, 45, 103
   masses of nuclei, 45
   odd-odd nuclei, 45
   powder model, 41
   quantum numbers, 43
   validity, range of, 46
Surface reactions, 89ff
Symmetry properties of nuclear states, 15, 36f, 43f
Systematics of stable nuclei, 12ff

T multiplets, 37
T quantum number, 36f
Teller, 98
Tensor interaction, 33
Thomson, J. J., 3

Threshold of reaction, 26
Two-body interactions, 32ff

Uniform model, 41ff

Watson, 61
Weak coupling, 85

Weisskopf, 87, 94
Weisskopf unit, 109, 110, 111
Weizsäcker, 15, 45, 108
Weizsäcker mass formula, 15
Widths of resonance levels, 27, 75, 78, 82

Yennie, 22